POLITICAL PHILOSOPHY NOW

Political Philosophy Now is a series which deals with authors, topics and periods in political philosophy from the perspective of their relevance to current debates. The series presents a spread of subjects and points of view from various traditions which include European and New World debates in political philosophy.

POLITICAL PHILOSOPHY NOW

How Kant Matters for Biology
A Philosophical History

Andrew Jones

UNIVERSITY OF WALES PRESS • 2023

www.uwp.co.uk

British Library Cataloguing-in-Publication Data
A catalogue record for this book is available from the British Library.

ISBN 978-1-78683-973-2
e-ISBN 978-1-78683-974-9

The right of Andrew Jones to be identified as author of this work has been asserted in accordance with sections 77, 78 and 79 of the Copyright, Designs and Patents Act 1988.

Typeset by Marie Doherty
Printed by CPI Antony Rowe, Melksham, United Kingdom

For my parents,
Paul, Barbara and Angela

Contents

Acknowledgements

This book originated from my PhD research. I thank the AHRC South West and Wales Doctoral Training Partnership for funding that research. I am indebted to Christopher Norris for his support and encouragement as I developed these arguments from their germinal forms. I also thank John Dupré for our invaluable discussions as my second supervisor. Building the foundations of this work under the guidance of two supervisors largely critical and suspicious of the Kantian standpoint was invaluable for this project. I also thank Angela Breitenbach and Jon Webber for their rigorous feedback during my PhD examination.

I owe special thanks to the commitment and rigor of the members of the Kant reading group that we formed at Cardiff University: Howard Williams, Andrew Vincent, John Saunders and Hugh Compston. None of us were prepared for the task we undertook when started our weekly meetings in 2015!

I am honoured that my postdoctoral research fellowship with the Theology and Religion department at the University of Exeter allowed me to develop my ideas into this book. My research, under the exceptional support and guidance of Christopher Southgate, forms part of the project 'God and the Book of Nature'. This project, funded by the John Templeton Foundation (Grant ID 61507), provided me with the exciting opportunity to extend my scope of research to the relationship between science and theology. Finally, I am especially thankful to the anonymous reviewer for this manuscript, their exceptional feedback was a guiding beacon in the later stages of its development. Any errors in this book are my own and its contents do not necessarily reflect the views of my funders.

Introduction

This is a book about understanding how philosophical ideas have influenced the development of science, with a specific focus on the case study of Immanuel Kant's influence on the development of biology in the British Isles. The central theme connecting the various arguments of this book is my conviction that re-examining Kant's historical influence on the development of biology can elucidate some of the predominant disputes in the contemporary philosophy of biology. This by no means entails that Kant is the only historical figure who might offer elucidation on these issues. My aim is to demonstrate how the history of philosophy and the philosophy of science can offer new perspectives on the issues facing contemporary philosophers.

The title of this book, *How Kant's Philosophy Matters for Biology*, draws from the structure of Kant's question in the *Critique of Pure Reason* – how are synthetic *a priori* propositions possible?.[1] Kant avoids delving into an enquiry concerning whether these judgements are possible, instead he begins his critical philosophy by asserting the undeniable existence of these judgements. I adopt a similar strategy in this book by focusing on how Kant matters, rather than questioning whether he matters, for biology. I build from previous accounts by philosophers and historians of biology that have focused on the significance of Kant's philosophy to investigate precisely how philosophers and scientists have deployed aspects of Kant's philosophy in their own theories. I argue that the driving force behind appeals to Kant often relate to core issues within biology that cannot be solved within the remit of naturalism alone. Moreover, the way scientists and philosophers have deployed Kantian principles for their own purposes often requires that these principles function in ways that are incompatible with their original functions in Kant's critical philosophy.

Kant's critical philosophy has a tenuous relationship with the philosophy of biology. Kant himself raised a number of problems regarding the possibility of biology being considered a proper natural science. A proper science must be grounded on rational principles that are apodictically certain and necessary, in contrast to the improper sciences that derive laws from contingent experience.[2] Even this statement should be approached with caution, as Kant's philosophy preceded, and was in part responsible for, the emergence of the biological sciences. The term 'biology' first appeared ten years after the publication of the third *Critique*, in 1800, in a footnote in a German medical journal. It then re-appeared independently in 1802 in the works of the German naturalist, Treviranus, and the French zoologist, Lamarck.[3]

There are various reasons to bring into question the compatibility between Kant's account of proper science and the driving principles behind contemporary science. The conditions of apodictic certainty and necessity that Kant outlined for the fulfilment of a proper science do not generally hold for contemporary science. Moreover, his account of science is inseparable from his broader critical project of transcendental idealism. Transcendental idealism establishes the conditions for distinguishing between knowledge of the appearance of objects as opposed to knowledge of objects in themselves. In contrast, philosophers of science generally do not argue for a difference in kind between the objects of experience and objects in themselves.

It might seem that the differences between the foundations of Kant's critical philosophy and philosophy of science are insurmountable, so that any collaboration between these two disciplines could not be fruitful. For instance, John Zammito has been critical of recent appeals to Kant's philosophy to help resolve issues in contemporary philosophy of biology.[4] However, research into the relationship between Kant and biology has been growing over the past thirty years. Kant's historical and philosophical importance for the development of biology is now widely acknowledged. In addition, certain contemporary philosophers of biology have appealed to aspects of Kant's discussion of teleological judgement in support of biological agency. These various attempts to understand the relationship between Kant and biology have overlapped at times, yet there has been little attempt to explore the benefits of combining these historical, philosophical and contemporary approaches toward Kant's significance for biology.

There is growing support for approaching issues in biology from interdisciplinary perspectives. Philosophers and historians can

often approach issues from alternative perspectives to those permitted by scientific methodology. Scott Lidgard and Lynn Nyhart
explain how these other disciplines can approach biological issues
from a different problem space, which includes critical engagement in the following areas: the context of the construction of
knowledge; the circulation of knowledge within and between communities; and changes in contexts and questions over time.[5] These
questions open up previously unexplored avenues of investigation,
allowing for a richer investigation of the development of scientific
theories.

In this book, I offer an examination of the historical and philosophical implications of the influence of Kant's critical philosophy
on the development of biology in the British Isles in the nineteenth
century. This dual approach makes it possible to consider the
extent to which Kant's influence on biology has been compatible
with the original deployment of his theory. Kant has a helpful distinction that helps to draw out this difference, I aim to distinguish
the justification (*quid juris*) from the fact (*quid facti*) of the matter.
The former concerns the entitlement or justification for the use of
a concept, while the latter concerns the fact that concepts are actually used.[6] When philosophers of science have engaged with Kant,
both historically and within contemporary philosophy of biology,
the subsequent developments on Kant's philosophy have rarely
been consistent with their original deployment in the context of
his critical philosophy. Hence, it is important to distinguish the *fact*
that these philosophers of biology have appealed to Kant's philosophy from their *justification* for doing so.

This is indicative of a deeper tension between transcendental idealism and naturalism, which has been the focus of hostility towards
the relevance of Kant's philosophy for contemporary philosophy of
biology. According to Zammito, '[n]aturalist philosophers of biology today need not succumb to the scruples (whether 'absolute'
or 'transcendental') that haunted eighteenth-century philosophers'.[7] Zammito's concern regarding the incompatibility between
naturalism and transcendentalism exposes an important difficulty
that arises when attempting to understand Kant's influence on the
development of biology. The tension between naturalism and transcendentalism is an important factor to consider when examining
Kant's influence on biology, but it cannot cause us to overlook crucial points where Kant's critical philosophy has been significant

both for the development of biology and for contemporary philosophers of biology.

Exploring these tensions exposes how certain principles in biology developed from, and continue to draw support from, ideas that do not easily fit with conventional understandings of naturalism; examples of which include biological autonomy and natural selection. I argue that both these ideas borrow from other ideas that help us to comprehend nature. Two ideas that are central to natural selection are the analogy between artificial and natural selection, and the similarity between organisms and machines. There is an indispensable role of judgement in our ability to apply these ideas to nature. The idea that the foundations of any aspect of biology are non-naturalistic is generally met with hostility. In contrast, I argue that this presents us with the opportunity not only to develop a richer account of the context of Kant's influence on the development of biology, but also to consider how Kant's philosophy can contribute to debates in contemporary philosophy of biology.

Previous accounts of Kant's influence on biology have focused on the development of nineteenth-century German biology. For instance, Timothy Lenoir argued that Kant's philosophy constituted the 'hardcore element' of the teleomechanist research programme. Zammito and Robert Richards have criticised Lenoir for not recognising the incompatibility between Kant's transcendental idealism and its subsequent naturalistic deployment within biology. By distinguishing questions pertaining to whether an influence is justified from the fact that an influence has occurred, it is possible to establish a middle ground between these two positions. I argue that the criticisms against Lenoir do not demonstrate that Kant did not influence biology, but rather that Kant's influence could not be sufficiently justified in accordance with biological naturalism. Denying that Kant could have influenced the development of biology, because this influence inherently misappropriates Kant's philosophy, places an unrealistic demand on the context of discovery of scientific theories. There are many instances where scientific theories develop in accordance with metaphors that stimulate research. According to Richard Lewontin, many of the foundations of science are metaphorical:

> It is not possible to do the work of science without using language that is filled with metaphors. Virtually the entire body of modern science is

an attempt to explain phenomena that cannot be experienced directly by human beings ... Physicists speak of 'waves' and 'particles' even though there is no medium in which the 'waves' move and no solidity to those 'particles'. Biologists speak of genes as 'blueprints' and DNA as 'information'.[8]

Many well-received aspects of science are inseparable from the contextual metaphors that have been instrumental for the development of scientific theories. This brings into question the assumption that science is grounded solely on naturalistic principles. For Lewontin, the cost of the use of such metaphors in science is that it must remain vigilant against confusing metaphors with the real things of interest. If we bestow on these metaphors a greater degree of reality than is warranted, we cease to see the world *as if* it is a certain a way and instead take it to *be* that way.[9] Lewontin identifies this as a central issue for philosophy of science; the separation of the metaphorical and real objects of science resonates with the motivation behind Kant's critical philosophy. Kant offered a reorientation of philosophical enquiry away from the view that objects independent of experience – or objects in themselves – have identical features to the objects of experience. Instead, Kant investigated the human faculties that make experience possible through his account of transcendental idealism. He approached our claims to knowledge in a juridical manner that focused on the justification for knowledge. By distinguishing the appropriate domains for knowledge, he aimed to provide the foundation for reason to 'secure its rightful claims while dismissing all its groundless pretensions, and this not by mere decrees but according to its own eternal and unchangeable laws; and this court is none other than the **critique of pure reason** itself'.[10]

Kant identified three faculties: sensibility; understanding; and reason. His critical philosophy, understood as an architectonic system, is an examination of how these faculties relate to one another to produce different forms of cognitive content that have different domains of application. Crucially, we must first understand the appropriate limits of knowledge, what Peter Strawson termed 'the bounds of sense', before we can engage in a scientific understanding of the world. Kant's account of teleological judgement (or biology) plays a crucial role in the architectonic structure of our faculties as finite rational beings. Kant's account of the organism is inseparable from the context of transcendental idealism. He aimed to develop a systematic account of the faculties of finite rational beings that establishes the necessary conditions for epistemology, metaphysics, morality, politics, teleological judgement, and more. Central to all these elements of his philosophy is the emphasis on the subject as the foundation for establishing synthetic *a priori* principles that reveal the necessary conditions for the possibility of knowledge in these diverse areas. The emphasis on identifying necessity within experience is one example of how aspects of transcendental idealism seem 'exactly backwards' or 'contrary to plain sense' according to contemporary conventional wisdom.[11]

The fundamental issue that this book aims to address – Kant's historical and philosophical relationship to biology – is inherited because both Kant scholars and philosophers of biology continue to appeal to Kant's critical philosophy for potential resolutions to issues in contemporary philosophy of biology. These appeals are unsurprising given that Kant influenced biology in important ways. However, these appeals create the opportunity to re-examine some fundamental assumptions about the development of biology. Some philosophers of biology have appealed to Kant's account of the organism to help understand the difficult relationship between biology and naturalism. These accounts often draw from aspects of Kant's account of the organism without contextualising their discussions within transcendental idealism. When considered in the context of transcendental idealism, Kant's philosophy offers critique, rather than support, to the naturalist assumptions of contemporary philosophy of biology. I consider how this Kant-inspired critique can help offer alternative perspectives to certain disputes within philosophy of biology.

This simultaneous concern with both historical accuracy and contemporary relevance is most evident in the structure of Chapters 2, 3 and 4. Each of these chapters begins with a historical examination of a specific issue followed by a discussion of the implications for contemporary philosophy of science. These include discussions around the status of scientific laws, the unity of science and accounts of the organism. Various aspects of the historical and philosophical investigation of Kant's influence on biology have far-reaching ramifications for contemporary debates. In this sense, this book contains various interconnected arguments intended to offer cumulative support for one another, but each chapter is also self-contained to the extent that it can be read in isolation from the others.

In Chapter 1 I survey previous debates on the topic of Kant's influence on the development of biology and demonstrate the need for an alternative account of influence. This chapter will be of particular interest to historians of science and philosophers of history. First, I consider various arguments from Strawson's interpretation of Kant, which reinvigorated Kant scholarship for the English-speaking world. I argue that aspects of Strawson's account disarms the critical force of Kant's philosophy for contemporary philosophy of science. Following this, I examine the change in conceptions of historical and scientific methodologies ushered in by Kuhn's account of scientific paradigms. Lenoir's argument that Kant influenced philosophy of biology in Germany builds on this change by drawing from Lakatos's theory of research programmes. He argued that Kant formed the hardcore element of the research programme that he terms 'teleomechanism'. Lenoir has been criticised for overlooking the incompatibility between Kant's account of organisms as the product of regulative judgement and the biological naturalistic understanding of organisms as entities that exist independent of judgement. I argue that this is a result of Lenoir's commitment to a Lakatosian account of influence, which entails that foundational principles, or the 'hardcore elements' of any research programme, are protected from critical enquiry. Instead, scientific experimentation is directed towards auxiliary hypotheses, that if proved wrong, do not bring the scientific theory into question. Lakatos's conception of research programmes was in part a response to Kuhn's account of scientific paradigms, which suggested that theories blindly accumulate anomalies without any method to judge their potential significance for the theory. I argue that both Kuhn's and Lakatos's accounts presuppose that an influence must be compatible with its source.

I contrast their accounts with alternative conceptions of influence that avoid this presupposition. I draw on aspects of the critical theorist Harold Bloom, the philosophers of science Paul Feyerabend and Angela Potochnik, and the philosopher of ideas Robin Collingwood. Bloom's account of poetic influence reveals how poets intentionally and creatively misrepresent the style of previous poets to forge their own identity through acts of 'misprision' or creative correction. Feyerabend shows how the development of new scientific theories is also often a creative process of discovery, which is then justified by subsequent experiments. Potochnik explores how science often uses models that make assumptions about the world, which are evidently false but nonetheless serve a crucial function for science as idealisations. She asserts that many metaphysical disputes are the result of transforming these idealisations from epistemic and pragmatic aspects of scientific practice into ontological principles. Collingwood's history of ideas suggests that metaphysical principles should be understood as historical absolute presuppositions that are incommensurable between different societies. Combined, these accounts provide evidence for the account of the development of scientific theories as a creative process, which utilises both natural and non-natural principles. This allows us to avoid the implicit requirement that Lenoir imposes on claims concerning Kant's influence on nineteenth-century German biology, specifically that influence presupposes compatibility. Zammito argues that Kant's influence on biology must have been minimal because biologists misunderstood his philosophy. In contrast, I argue that influence needs to include cases where philosophers have been misunderstood, as it creates new perspectives for the philosophers and issues under examination. There is much to be learnt from investigating the precise nature of these misunderstandings and their subsequent impact on the development of philosophy and philosophy of science. My account of influence allows us to appreciate Kant's influence on the development of biology without demanding that Kant's critical philosophy is compatible with biology.

Chapter 2 will be of the most interest to Kant scholars, as I argue that fundamental aspects of Kant's theoretical and practical philosophy were motivated by David Hume's impact on his intellectual development. Examining Kant's philosophy as developing from Hume shows how transcendental idealism is in part responding to the shortcomings of scepticism. Kant recognised a parity between Hume's philosophy and his own critical philosophy because of their shared concern with the possibility of deriving certainty from appearances. Importantly, Kant misidentifies this parity with Hume's matters of fact because Kant misunderstood Hume's distinction between matters of facts and relations of ideas since he regarded it as a logical, rather than a psychological, distinction. For Hume, relations of ideas were not truly independent of experience, but rather that they required a single experience to demonstrate their certainty. This exposes a strong similarity between Hume's relations of ideas and Kant's account of synthetic *a priori* truths, which Kant overlooked.

My interpretation of the relationship between Hume and Kant portrays Kant as developing from the shortcomings of Hume's sceptical empiricism and subsequently revealing how some of the fundamental tenets of transcendental idealism offer a response to Hume. Kant praised Hume for demonstrating the impossibility of establishing a rational ground for knowledge of necessary causal connections relating to objects in themselves. He regarded transcendental idealism as developing from the problems identified by Hume's naturalism. My interpretation is significant for three distinct areas of scholarship: Hume studies; Kant studies; and contemporary accounts of laws in philosophy of science.

I contrast my interpretation with the sceptical realist interpretation of Hume's philosophy, which proposes that Hume believed in the existence of metaphysical laws despite lacking sufficient evidence to verify this belief. Considering Kant's philosophy as a response to Hume's scepticism offers the foundations for strong criticism against the sceptical realist interpretation. Even if Hume believed in the existence of metaphysical laws, Kant demonstrates how searching for metaphysical laws beyond experience is meaningless given Hume's self-imposed philosophical restrictions. Moreover, I argue that Kant's restriction of knowledge of theoretical laws to the necessary conditions for the possibility of experience is consistent with his broader response to Hume.

Having defined Kant's critical turn in relation to Hume, I then apply my interpretation of Kant to recent accounts of the metaphysical status of laws in philosophy of science. I focus on Roy Bhaskar's transcendental (or critical) realism and Nancy Cartwright's nomological pluralism. The fundamental difference between these accounts is that Bhaskar argues that we must presuppose the existence of unchanging (or intransitive) necessary laws for the possibility of science, while according to Cartwright there is little empirical justification for the existence of such laws. Both present their positions in the form of transcendental arguments, yet they arrive at incompatible conclusions. From a Kantian perspective, it is impossible for valid transcendental arguments to arrive at incompatible conclusions. I argue that the underlying reason for these incompatible conclusions is that both Bhaskar and Cartwright attempt to use transcendental arguments to derive necessary conditions about metaphysical possibility, rather than conditions for the possibility of experience. I argue that their arguments can be presented as a version of a Kantian mathematical antinomy; combined they reveal that knowledge of scientific laws can never extend to entities independent of perception.

In Chapter 3, I examine Kant's influence on the development of nineteenth-century biology in the British Isles. This chapter will be of particular interest to philosophers and historians of science. I argue that the philosopher of science, William Whewell, derived his account of the active powers of the mind from Kant. Kant's influence, I suggest, was double-edged: Kant helped Whewell to consider the active powers of the mind in its engagement with nature; however, these active powers were inseparable from transcendental idealism in the context of Kant's critical philosophy. Whewell attempted to separate the importance of the active powers of the mind from the broader implications of transcendental idealism. He regarded knowledge as the product of a fundamental antithesis between thoughts and things. For Whewell, the ability for science to develop theories that explained a broader scope of phenomena under general principles (or what he termed 'consilience') was sufficient to conclude that those scientific theories were discovering facts about the real nature of objects in themselves. In contrast, Kant denied any possibility of knowledge of these objects. For Whewell, a philosophy for scientific method must account for the unpredictable ways that scientific knowledge arises. These scientific discoveries could not be adequately explained as scientific truths if constitutive knowledge was restricted to Kant's categories of the understanding. He transformed Kant's philosophy by asserting that the difference between constitutive and regulative principles differed only in degree, not in kind. In contrast to Kant, he argues that regulative judgements can become constitutive. Whewell's transformation of Kant's philosophy was possible only because Whewell stipulated that the existence of God made it possible for us to know when our scientific theories correspond to a world independent of our experience. Whewell's appeal to God is problematic from the perspective of transcendental idealism because Kant argued that knowledge of God's existence was beyond the remit of theoretical knowledge. The relationship between theology and the principle of consilience gains significance as it has remained an important methodological principle for scientific practice.

My comparison of Kant and Whewell provides a foundation for critically examining some contemporary Kant scholars who argue in support of the benefits of incorporating Kantian principles into scientific methodology. I argue that these accounts require us to overlook incompatibilities between Kant's transcendental idealism and contemporary philosophy of science.

In Chapter 4, I assess the extent to which Whewell influenced Charles Darwin. I argue that the two significant influences on the methodology of Darwin's theory of natural selection were John Herschel and Whewell. Whewell's primary influence on Darwin, I suggest, was Darwin's commitment to the principle of consilience. Darwin emphasised the capacity of the theory of natural selection to provide a unified explanation of the origin of species. Whewell's account of biological entities, which was influenced by aspects of Kant's philosophy, was less significant on the development of Darwin's account. Whewell transformed Kant's discussion of teleological judgement into justification for the ability to perceive God's divine plan. Whewell's transformation of Kant's account meant that Darwin remained unaware of the aspects of Kant's account of teleological judgement that conflicted with his account, such as the analogy between artificial and natural selection. Darwin deployed this analogy in part because of the influence of William Paley's theological explanation of the apparent design of organisms. Paley argued that the design of organisms was analogous to finding a watch on a heath. Kant denied this analogy between artefacts and organisms on the basis that organisms are judged to possess the capacity for self-organisation of their parts and wholes, whereas artefacts lack this capacity and depend on an external source for their organisation.

I examine the recent debates that suggest Kant was a significant figure for understanding Darwin's theory of natural selection. For Michael Ruse, Darwin was fundamentally committed to a mechanistic account of nature, yet he argues that Darwin considered nature *as if* it were the product of design. He argues that Darwin would have agreed with Kant's conception of teleological judgement as a heuristic principle. His appeal to Kant overlooks significant incompatibilities between the accounts of Kant and

Darwin. For Kant, the possibility of the conception of organisms depends on our ability to judge nature teleologically, it is not heuristic in the sense that we *choose* to apply such judgements to experiences of organisms that enable us to examine them *as if* they were designed. Rather, Kant argued that teleology was a necessary precondition for the possibility of experiencing organisms.

From this discussion of Kant and Darwin, I consider how the status of design has changed in contemporary philosophy of biology. Stephen J. Gould and Lewontin criticised adaptationism because of its underlying assumption that the purpose a trait currently serves is causally responsible for the emergence of that trait. Daniel Dennett's adaptationist account also requires a Kantian non-naturalistic foundation to make it possible to conceive of nature *as if* it were intentionally designed. In contrast, other philosophers of biology have argued that organisms do not need to be understood in accordance with design. It has been argued that organisms should not be understood as being similar to machines, rather instead that they share more similarities with non-organismic dissipative structures such as whirlpools. Examining design from a Kantian perspective allows us to understand how judging nature in accordance with design allows us to distinguish clearly between organisms and non-organismic systems.

In Chapter 5, I build from the previous chapters to demonstrate how these Kant-inspired criticisms of biology can also be developed into a source of guidance and support biology. I argue that certain aspects of Kant's political philosophy can guide scientific enquiry at the societal level because biology cannot be justified in accordance with naturalism alone. The chapter begins with a discussion of how the conditions for demarcating biological individuals persists as a controversy for contemporary philosophy of biology. I discuss three specific cases that demonstrate problems inherent to establishing a single definition of biological individuality: genetic homogeneity; symbiosis; and niche construction. Taken in combination, these cases demonstrate how the conditions used to define biological individuals demarcate living and non-living entities differently and can come into conflict with one another. The difficulties associated with offering a single explanation for the demarcation of biological entities offers support to my proposal that we should consider the benefits of revisiting Kant's demarcation of biological individuals as inextricably connected to judgement.

Following this, I examine the predominant appeals to Kant by contemporary philosophers of biology. Some leading proponents of biological autonomy appeal to Kant's discussion of teleological judgement as offering support to their account. These biologists argue that Kant identified how organisms possess the properties of self-organisation, which can be explained as part of nature. This identification of Kant with naturalism overlooks significant tensions between naturalism and Kant's philosophy. He argued that we only judge nature *as if* it is the product of final causes, and therefore our ability to perceive certain objects as self-organising is not entirely naturalistic as it is dependent on judgement. The appeal to Kant by proponents of biological autonomy overlooks the broader relationship that Kant's account of the organism has with his accounts of moral and physical teleology. Kant thought that our ability to conceive of organisms in nature was inseparable from certain features of our capacity for practical reasons. On the one side, the inseparability of teleology from Kant's broader philosophy raises concerns about the suitability of Kant for the aims of contemporary theories of biological autonomy. On the other hand, these appeals to Kant potentially expose the problematic foundations of contemporary accounts of biologically autonomous systems.

To explore how the connection between moral and teleological judgement might have positive implications for contemporary philosophy of biology, I draw from aspects of Kant's political philosophy to demonstrate his relevance for developing a human-centred approach to scientific research. John Dupré indicates that his account of biological freedom is minimally compatible with Kant's morality on the basis that both regard the subject as the source of action. He argues that collective human action has been a source of major evolutionary change that is unexplainable according to a reductive genetic account of biology. Dupré's examples generally focus on collective human actions that have been beneficial for society, such as widespread access to healthcare and education; however, I argue that his account cannot provide the basis for promoting these beneficial human behaviours over detrimental ones. Elements of Kant's political philosophy offers a possible framework for recognising the importance of collective human action as a source of evolutionary change. Moreover, he also recognises the importance of developing an account of social responsibility aimed at promoting beneficial and progressive collective human actions. Of course, there is no clear way to demarcate between those actions that are beneficial for society from those that are not. Nevertheless, Kant's reservations about revolution and war and his arguments in favour of the state's responsibility towards the minimal needs of all citizens, are indicative of his aim to identify the conditions for a progressive political philosophy.

By overlooking the importance of moral teleology for physical teleology within Kant's philosophy, contemporary appeals to Kant do not consider how he offers an account of directing scientific research towards issues that have positive broader societal implications. Kant never considered the societal importance of science because of the strict conditions that he imposed on what constituted a proper science, conditions that importantly excluded biology from becoming a proper science. Scientific developments have demonstrated repeatedly that the Newtonian framework of science, which Kant held as the paradigmatic example of proper science, is an exception to the general development of science. The complex relationship between science and society is gaining recognition in contemporary philosophy. I conclude the book by arguing that Kant's political philosophy offers important insights on the need for an explicit explanation of the social responsibilities of science to engage in research that is directed towards the collective societal good.

In summary, I develop a new perspective to debates relating to Kant's significance to biology by proposing a broader conception of influence. Scientists and philosophers have appealed to aspects of Kant's philosophy to solve issues in biology without appreciating that these principles can be justified only in accordance with transcendental idealism. More generally, this book offers a critical analysis of the status of naturalism within certain aspects of biology. I argue that at various points in the development of biology, biologists have appealed to principles that are incompatible with naturalism. I examine the broader implications of these appeals for their accounts and re-evaluate the importance of broadening our understanding of the biological sciences by incorporating factors and principles that reach beyond the remit of naturalism.

1 • Understanding Influence: the Role of Transcendental Idealism for the Development of Biology

Introduction

Kant's influence on the development of biology has been presented in conflicting ways. As the frequent references to Kant in histories of biology attest, his philosophy was important for the formation of a biological science. However, biologists did not keep their accounts consistent with the limitations imposed on knowledge by Kant's transcendental idealism. In this chapter, I argue that his influence is best understood as being based on a misunderstanding of transcendental idealism on the part of biologists. This does not mean that transcendental idealism is useless for biology, as some have argued; instead, we can understand Kant's complex relationship with biology as centring around problems emerging from the tendency within biology to overstep the commitment to naturalism.[1] This is a deeply rooted issue that first emerged within Kant's own lifetime, which is evident from his engagement with Johann Friedrich Blumenbach's account of the formative force (*Bildungstrieb*). Blumenbach regarded the *Bildungstrieb* as a law of nature of the same calibre that Newton had discovered for physics, whereas Kant regarded our ability to view nature in accordance with teleological causes as belonging to our form of judgement. All we can say is that it is a condition for the possibility of biology that we judge nature *as if* it is governed by teleological principles. Kant regarded the inseparability of biology from our judgement as sufficient reason to deny that biology could become a science; however, there is growing support within contemporary philosophy of biology for positions that recognise the inherent human-centric nature of science. This offers an alternative explanation for Kant's relevance for contemporary philosophers of biology as Kant regarded biology as inseparability from human judgement.

Discussions of Kant's importance for the historical development of biology are indebted to Lenoir. He argued that the development of nineteenth-century German biology emerged from Kant's account of teleological judgement in the third *Critique*. Lenoir's account has been subject to criticism by Zammito,[2] who responded by highlighting how Lenoir exaggerates Kant's influence because the fundamental differences between transcendental idealism and naturalism are insurmountable; biology is grounded on the principles of naturalism 'and Kant simply cannot be refashioned into a naturalist'.[3] By examining the conceptions of influence held by Zammito and Lenoir, I argue that both impose a strong level of similitude as a precondition for influence. The result is that neither can adequately account for how influence can arise from a misunderstanding. This leads Zammito to conclude that Kant was of little significance for the development of German biology, whereas Lenoir plays down the differences arising from the incompatibilities between Kant and the members of the teleomechanist research programme. Lenoir's description of teleomechanism draws from the philosopher of science, Imre Lakatos, specifically his conception of research programmes. I argue that Lakatos's account is not fit for purpose regarding Lenoir's account because it cannot distinguish between influences that are based on misunderstanding and those that are not.

The investigation of Kant's influence on biology should be distinguished from questions pertaining to the compatibility of Kant's account with those that it may have influenced. I intend to show that the existing debate about Kant's influence on biology conflates these questions. Like Kant's first *Critique*, this chapter can be viewed as a ground-clearing exercise,[4] which makes it possible to expand on the notion of influence in a way that appreciates the importance of misunderstandings. I develop an alternative explanation of Kant's influence on biology that recognises a tension between transcendental idealism and biological naturalism, while maintaining that an influence has nonetheless occurred.

1.1 Transcendental idealism as subservient to the scientifically minded philosopher

Over the past half a century, Peter Strawson's interpretation of Kant has played a significant role in the rejuvenation of interest in

Kantian philosophy in the English-speaking world. However, his account significantly reinterprets aspects of Kant's philosophy to focus on issues that are relevant to philosophical issues of his time. In this section, I offer a defence for transcendental idealism against several aspects of Strawson's interpretation of Kant. First, I contrast Strawson's criticism of Kant's separation of appearances from things in themselves with Lucy Allais's interpretation of Kant as a moderate metaphysical realist. It follows from Allais's account that Kant's distinction between appearances and things in themselves is a point of similarity, not divergence, between Kant and contemporary philosophers of science. Second, Strawson's rejection of Kant's account of the self as the transcendental unity of apperception disarms an important argument that emerges from transcendental idealism; namely, that we cannot deduce the origin of the transcendental conditions required for knowledge either physiologically or biologically. Strawson's interpretation of Kant is unable to identify these aspects of his philosophy because it is inherently limiting to examine Kant from the perspective of what he terms a 'scientifically minded philosopher'. I argue that Strawson's scientifically minded philosopher is a caricature that no longer corresponds with many of the principles generally held by contemporary philosophers of science.

1.1.1 The scientifically minded philosopher and the relation between appearances and things in themselves

Kant's denial of the correspondence between appearances and things in themselves is often cited as a source of disparity between transcendental idealism and more recent accounts of philosophy of science. Strawson takes issue with the apparent incompatibility between Kant's philosophy and our best scientific understanding at the time of the publication of his *The Bounds of Sense*, originally published in 1966. By outlining this objection and Kant's potential response, I attempt to uncover some of the fundamental tenets of Kant's philosophy. Following this, I examine Allais's interpretation of Kant as a moderate metaphysical realist to reveal that these aspects of Kant's philosophy offer a potential compatibility between Kant and the sciences. Let us begin with Strawson's comparison between Kant and the scientifically minded philosopher:

The scientifically minded philosopher does not deny empirical knowledge of those things, as they are in themselves, which affect us to
produce sensible appearances. He only denies that the properties which,
under normal conditions, those things sensibly appear to us to have are
included (or are *all* included) among the properties which they have,
and which we know them to have, as they are in themselves. But Kant
denies the possibility of any empirical knowledge at all of those things,
as they are in themselves, which affect us to produce sensible experience. It is evidently consistent with, indeed required by, this denial to
deny the physical objects of science *are* those things, as they are in
themselves, which affect us to produce sensible experience.[5]

The transcendental idealist cannot assume that our scientific knowledge relates to properties of the objects 'as they are in
themselves'. Strawson correctly identifies that transcendental idealists must reject the possibility that objects of science are things
in themselves. Kant denies our experience relates to such objects
because experience is the product of the faculties of sensibility and
understanding. Space and time are not properties of objects as they
are in themselves, rather they belong to the faculty of sensibility.
Strawson claims that transcendental idealists must re-evaluate
their commitment to space and time as generated by the faculty of
sensibility. If they do not, then they are committed to the position
that things as they really are, in themselves, are not in space and
time. Strawson outright denies this possibility: '[t]hings, as they
really are, are not removed from the spatio-temporal framework of
reference. They are simply things as science speaks of them rather
than as we perceive them.'[6]
Strawson's claim of the incompatibility between the ways that
scientists and Kant view space and time is strongly connected to
his broader criticism against Kant, referred to as 'the problem of
affection'. The problem of affection refers to Kant's denial that we
can know anything about the origin or cause of experience because
we can only know the content of experience. Strawson regarded
scientific developments since Kant's lifetime to have provided sufficient evidence to relegate this aspect of transcendental idealism to
absurdity. Strawson did not expand on the specific nature of these
developments, but for him they seemed to demonstrate that the
thing in itself was undeniable because it was the object of scientific
enquiry.

Strawson's arguments that Kant is incompatible with science because of his separation of appearances from things in themselves is a caricature of Kant's philosophy. Importantly, Kant did not regard the unknowability of the thing in itself as a fundamental problem as the thing in itself 'is never asked after in experience'.[7] The origin of experience – as an object independent of experience – is not given in experience. It is not possible to enquire into the origin of our faculties because this would require us to presuppose that space and time existed independently from the faculty of sensibility.

For Kant, space and time are both empirically real and transcendentally ideal. Any object of experience is empirically real insofar as it is an appearance, but we cannot provide an explanation of any empirical object that goes beyond what is given in appearance: 'it is nothing at all if one abstracts from the subjective conditions of sensible intuition.'[8] Strawson disregards this aspect of transcendental idealism when he explains experience as causally dependent on the relation between our physiology and things in themselves. We are primarily creatures, as opposed to finite rational beings, that are already *in* time and space as we each have a history and a bulk.[9]

In contrast to Strawson, Lucy Allais's interpretation offers a more sympathetic account of both Kant's rejection of knowledge of objects in themselves and the need for such knowledge in contemporary philosophy of science. Allais describes her interpretation as a 'moderate metaphysical realism'. According to Allais, 'the things of which we have knowledge have a way they are in themselves that is not cognizable by us ... the appearances of these things are genuinely mind-dependent, while not existing merely in the mind'.[10] Importantly, she argues that this aspect of Kant's philosophy is potentially compatible with contemporary philosophy of science. The contents of science are generally directed towards an empirically observable reality. Allais interprets Kant as saying that 'the things of which we have experience are the things that are described by science, what we know through science counts as part of possible experience, and neither science nor perceptual experience gives us knowledge of an unobservable reality'.[11]

Allais reverses Strawson's charge that Kant's philosophy is counterintuitive because it denies knowledge of things in themselves. Her argument is that the aims of science and Kant's theoretical philosophy are compatible insofar as neither is concerned with demonstrating the existence of entities in a reality beyond possible observation. Therefore, Kant's rejection of knowledge of things in themselves is consistent with contemporary philosophy of science to some extent. Moreover, this reversal means that it is more counterintuitive for scientific justification to appeal to anything that is essentially unobservable.

There is widespread agreement that science should relate to observable experience, but Allais's interpretation entails the stronger stance that science lacks the justification to make claims about anything beyond observation. In this context, the relation between science and unobservable reality can be understood in two ways; either science *does* not talk about unobservable reality, or it *should* not. The former would be problematic as many contemporary debates in the philosophy of science discuss unobservable entities. For instance, consider the debate in philosophy of biology between the opposed positions of metaphysical monism and metaphysical pluralism. For the philosopher of science, John Dupré, scientific monism is committed to metaphysical assumptions about the underlying scientific unity of nature, which lacks sufficient empirical grounds.[12] In contrast, he argues that there is greater empirical support for conceiving of nature as ontological processes, which can be more appropriately explained in accordance with metaphysical pluralism. According to Dupré and Daniel Nicholson:

> metaphysics is generally to be established through empirical means, and is ultimately therefore answerable to epistemology. Scientific and metaphysical conclusions do not differ in kind, or in the sorts of arguments that can be given for them, but in their degree of generality and abstraction.[13]

It is important to emphasise that this definition of metaphysics is significantly different from traditional conceptions. Aristotle defined metaphysics as 'first philosophy', or the study of 'being *qua* being'. This separated metaphysics from experience. In contrast, philosophers of science such as Dupré approach metaphysics as inherently related to empiricism; in this context, it could more appropriately be termed 'being *qua* experience'. This emphasis on the connection between experience and metaphysics does not entail that contemporary philosophers of science endorse Kant's denial of knowledge of entities independent of perception. Dupré and Nicholson argue that they are compelled to understand the metaphysical constitution of the world more generally in terms of universal processualism. However, they concede that the evidence for such a view can only be developed in a 'piecemeal' way.

This raises some important difficulties when considering the potential relationship between Kant and contemporary philosophy of biology. On the surface, it seems that the emphasis on experience as a condition of knowledge indicates a potential compatibility between their philosophies. Yet, the relation between experience and metaphysics is significantly different for each. Kant was critical of the idea that metaphysics could develop in accordance with anything other than a systematic method. Despite the growing empirical evidence for process biology and pluralism, Kant would deny any metaphysical account that developed in a piecemeal or fragmentary manner. This connection between scientific method and systematicity is a recurring theme throughout Kant's critical philosophy. It is most commonly associated with the *Metaphysical Foundations of Natural Science* but is also evident in his later political philosophy. He begins his 'Metaphysical Principles of the Doctrine of Virtue' with the following definition of metaphysics:

A philosophy of any subject (a system of rational cognition from concepts) requires a system of pure rational concepts independent of any conditions of intuition, that is, a *metaphysics* ... so that it can be set forth as a genuine science (systematically) and not merely as an aggregate of precepts sought out one by one (fragmentarily).[14]

Kant's hostility towards developing a metaphysics in a fragmentary or piecemeal manner creates significant tension between his critical philosophy and the growing support for pluralism in contemporary philosophy of science. The condition of systematicity that Kant imposes on scientific method is more restrictive than many contemporary philosophers of science would concede.

1.1.2 Strawson's account of the self

Another important aspect of Strawson's interpretation of Kant is his treatment of Kant's conception of the self. Strawson regards

Kant's transcendental unity of apperception physiologically, rather than transcendentally. He asserts that 'self-consciousness must ... belong to the history of ... a being which *has* a history and hence is not a supersensible being'.[15] In contrast, Kant argued that the self-consciousness of the unity of apperception was not historical, but rather a transcendental presupposition for the possibility of consciousness as belonging to a self. The physiological explanation of the self makes an epistemic leap that cannot be justified according to the principles of transcendental idealism. Kant emphasises that 'consciousness of the self is very far from being a knowledge[16] of the self'.[17] Explaining the self from the physiological perspective limits our knowledge of the self to a historical examination of its physiological emergence.

This tension between transcendental and physiological explanations exposes a difficulty for explaining Kant's potential influence on biology. Kant argued that the transcendental conditions of experience could not be derived physiologically, yet he also drew from biological ideas of his time to explain transcendental idealism. He developed an explanation of the faculty of pure reason that did not implicate reason as emerging from either God or nature. The biological notion of epigenesis was instrumental for his explanation. However, in its biological context, epigenesis could not offer an explanation for how the system of pure reason could emerge without itself being part of nature.[18] Hence, Kant argued that his appeal to epigenesis was not concerned with nature but with the epigenetic emergence of the system of pure reason.[19]

According to Jennifer Mensch, '[o]nly "the epigenesis of reason," as appealing neither to experience nor to God but only to itself, could finally serve as the ground for experience'.[20] From Strawson's account, we might assume that Kant was not aware how physiological explanations could now replace transcendental explanations because they fit more appropriately with our best scientific understanding. Kant was well aware of physiological explanations, and his response against such explanations provides crucial insight to his account of the limits of our knowledge. A physiological derivation of the self fails to grasp the properties that are specific to our capacity for reason. In his criticism of John Locke, Kant argued against any possibility of deriving the rational features specific to transcendental idealism physiologically:

I therefore call this attempted physiological derivation, which cannot properly be called a deduction at all because it concerns a *quaestio facti*, the explanation of the **possession** of pure cognition. It is therefore clear that only a transcendental and never an empirical deduction of them can be given, and that in regard to pure *a priori* concepts empirical deductions are nothing but idle attempts, which can occupy only those who have not grasped the entirely distinctive nature of these cognitions.[21]

Strawson assumes that the scientifically minded philosopher has achieved what Kant thought was impossible; namely, that physiological or biological explanations could take the place of a transcendental one. However, physiological and biological explanations are not as certain as Strawson suggests. The limit of biological explanations is still an important issue for contemporary philosophers of biology. According to Stuart Kauffman, 'there may be a limit to the way Newton taught us to do science and a need to reformulate what we do when we and other agents get on with living a life ... [I]t appears something profound is going on in the universe that is not finitely prestatable'.[22] The realisation that there might be a possible limit, or an unrealistic expectation, for biological explanation resonates with Kant. Kant also denied any possibility that 'there may yet arise a Newton who could make comprehensible even the generation of a blade of grass according to natural laws that no intention has ordered'.[23] Kant addresses the limitations of biological explanations in the second part of his third *Critique*; however, his conclusions about biology are inherently connected to other elements of his critical philosophy. In the context of the first *Critique*, the appeal to the epigenesis of reason reinforced the barrier between reason and nature as it suggests that we cannot understand reason as emerging from nature, which Kant returns to in the third *Critique*. According to Mensch, '[e]pigenesis thus served as a resource for a *metaphysical* portrait of reason, even as it was denied determinate efficacy in the world of organisms'.[24] There is an incompatibility between the accounts of the organism according to transcendental idealism and biological naturalism. It will be crucial to address this incompatibility to understand Kant's potential influence on the development of biology.

Again, Strawson prematurely assumes that scientific developments since Kant have irrefutably demonstrated that his transcendental account of the self is false. However, Kant's rejection of the physiological account of the self is not a mere historical over-sight of his critical philosophy that is now irrelevant due to scientific developments. For Kant, physiology is not sufficient to explain the development of all the aspects of humanity. Kant proclaims that the history of humanity is 'to be found neither in metaphysics nor in a museum of natural history in which the skeleton of man can be compared with that of other kinds of animal'.[25]

1.1.3 Strawson and the problem of translation, transcendental idealism and transcendental realism

Thomas Kuhn's account of the untranslatability between scien-tific paradigms helps to illustrate the shortcomings of Strawson's interpretation of Kant. Kuhn describes how the history of science essentially comprises independent scientific theories or paradigms that cannot engage with one another on their own terms. Kuhn uses the connotations of translation to explain this: '[t]o translate a theory or worldview into one's own language is not to make it one's own. For that one must go native, discover that one is think-ing and working in, not simply translating out of, a language that was previously foreign.'[26] Strawson simply does not approach Kant's philosophy as a native. Kuhn explains that the largest obs-tacle for historians of science is to appreciate the context under which the works that they examine are produced. Any histori-cal understanding must attempt to engage with the ideas under examination as a native, even if this is not ultimately achievable. Strawson's approach regards the presuppositions of the 'scien-tifically minded philosopher' as a lens for identifying the relevant features of Kant's work. According to Henry Allison:

> Strawson's way of formulating Kant's argument is fundamentally misleading. Kant is not maintaining that because we don't know 'real' objects (things in themselves) we have to make do with some kind of sub-jective 'Surrogate', but simply that the basis upon which the mind judges, the ground of its recognition of an objective world as distinct from itself is ultimately one of coherence. Rather than a 'Surrogate' he is offering us a transcendental re-interpretation of what is meant by the real.[27]

Strawson transforms transcendental idealism into a transcend-ent metaphysical position. For Allison, Strawson's transcendent metaphysical approach to Kant's philosophy meant that he did not allow himself to understand the nuances of transcendental idealism. Allison describes this as a 'transcendent conception of the transcendental'.[28] There is an important difference between Kant's argument that the objective world relates to appearance and his argument that we cannot have knowledge about things in themselves. Strawson's account interprets Kant's account as arguing that we do not have objective knowledge, because objective knowledge can only apply to things in themselves.

Kant contrasted transcendental idealism with transcendental realism. For transcendental realists, space and time are not merely conditions of phenomenal experience; rather they are real features of objects. They regard our representations of objects to pertain to objects in themselves because they apply the 'modifications of our sensibility into things subsisting in themselves'.[29] Transcendental realists make transcendent claims, which cannot be supported by reference to any experience whatsoever. Their claims are beyond any possible empirical justification and, therefore, Kant argues that these are dogmatic claims. When metaphysical disputes arise between transcendental realists, neither side can justify their position with empirical support. For Kant, the contents of their speculations are merely machinations that arise out of private vanity[30] with no regard for truth. Disputes between transcendental realists result in contradictions because they rely on assertions concerning states of affairs 'which neither of them can exhibit in an actual or even possible experience'.[31]

Transcendental realism is concerned with questions about being that are beyond the boundaries of knowledge, whereas transcendental idealism is concerned with identifying the conditions that allow us to justify our knowledge claims. Allison succinctly explains this difference: '[i]nstead of the traditional concern of metaphysics with the nature of being, it [i.e., transcendental idealism] advocates the primacy of the concern with the conditions of our knowledge of being'.[32] Any philosophy that does not acknowledge the discursive foundations of knowledge is committed to some version of transcendental realism, 'which is to say every philosophy except transcendental idealism'.[33] Allison argues that the number of philosophies included under transcendental realism, many of which would not be traditionally understood as 'realisms' at all, suggests that Kant did not regard transcendental realism as a metaphysical thesis.

According to Allison, 'if transcendental realism is not a metaphysical thesis then neither is transcendental idealism, since Kant presents them as contradictory opposites'.[34] Clearly, specific transcendent realist accounts are committed to certain metaphysical theses, yet we should not understand transcendental realism in general as a single metaphysical thesis because alternative transcendental realist accounts can contradict one another. Their commonality is that they all share a commitment to the idea that the conditions of experience are also conditions of objects as they are in themselves.

In contrast, the methodological aim of transcendental idealism is to expose the conditions of knowledge. It is a method, rather than a metaphysics, which focuses on the implications of specific orientations in thinking. Kant highlighted how his critical philosophy made it possible to expose the limitations of alternative philosophies by means of their orientation. Kant asserts that '[o]ne remains safe from all error if one does not undertake to judge where one does not know what is required for a determinate judgment'.[35] Transcendental idealism, understood as an orientation or methodology, is not metaphysical in the same way as specific transcendental realisms because it does not advocate a transcendent metaphysics.

Strawson's account prioritises transcendent metaphysical claims over transcendental idealism because of his emphasis on physiological explanations. Physiological explanations require metaphysical assumptions that are incompatible with transcendental idealism; namely, that space and time are prior to experience as conditions of physiological development. Transcendental idealism is not metaphysical because it opposes the notion that we can attain knowledge of entities independent of experience.[36]

Strawson reduces Kant to a figure who deserves respect because of his achievements despite working under the constraints of an inherently flawed system. He concludes, '[t]hat he conducted the operation under self-imposed handicaps – though not in itself a matter for congratulation – makes it more remarkable that he achieved so much'.[37] This compliment is double-edged, his achievements are only worthy of congratulations under the context of the limitations of transcendental idealism. Strawson does not consider how transcendental idealism, along with its apparent handicaps, exposes important metaphysical assumptions that are central to Strawson's account.

1.2 The role of influence and theory for history and science

The accounts of the relationship between influence and theory have significantly changed within philosophy of science. In 1960, Isaiah Berlin distinguished between the applications of theory in science and history and surveyed the differences regarding each of these methodologies. He regarded historical methodology as deficient to scientific methodology in various ways. Thomas Kuhn's account of scientific revolutions marked a significant shift in understanding the difference between science and history. Kuhn demonstrated how science does not culminate in a single rational theory as Berlin had proposed. Rather, the development of science is a series of historical paradigm shifts or revolutions in science whereby one theory overthrows another.[38]

1.2.1 Berlin and Kuhn on the relation between history and science

Berlin compared the methodologies of history and science and argued that history could never become a science. Science alone could provide a single correct account of reality governed by rational principles. In his article 'History and Theory: the Concept of Scientific History', Berlin presents the account of science that was predominant leading up to the publication of Kuhn's *The Structure of Scientific Revolutions*, which brought into question many aspects of scientific methodology that were taken for granted by Berlin.

Berlin regarded historical methodology as deficient to scientific methodology, as science exemplified a rational and logical structure that history lacked; 'natural science is nothing if it is not a systematic interlacing of theories and doctrines, built up ... [b]y the most competent practitioners in the field'.[39] The fundamental difference between these methodologies related to a difference in their treatments of facts in relation to theories. Historians regard facts as having a greater refutational power over their theories than scientists. Berlin argues that if a scientist had attempted to watch the sunrise, but the sun did not rise as had been expected, then it would be premature for the scientist to doubt the succession of day and night, or even our entire understanding of celestial mechanics, based on this experience alone. Scientists tend to look for additional reasons (or auxiliary hypotheses) that, if proven correct by additional experimentation, would account for the failure of the expectation of this experiment while remaining consistent with the theory overall.

In contrast, historical methodology does not place the same level of certainty on the theories that it prescribes. Consider the following example: a historian has a theory that forbids French generals to go into battle wearing a three-cornered hat, but the historian also possessed evidence that 'Napoleon had been seen in a three-cornered hat at a given moment during the battle of Austerlitz'.[40] Berlin argues that the historian should not continue to believe their theory considering this evidence: 'addiction to theory – being doctrinaire – is a term of abuse applied to historians; it is not an insult if applied to a natural scientist.'[41] He argues that history is not scientific because it does not approach nature systematically. Berlin described history as a 'skill' or an 'empirical knack' because the success of the historical account depends upon the judgement and skill of that historian. In contrast, the success of a scientist is not dictated by skill, or even intelligence, as they rely on theories. According to Berlin, '[a] man who lacks common intelligence can be a physicist of genius, but not even a mediocre historian'.[42]

Kuhn's examination of the history of science provided an alternative perspective on these aspects of science. He argued that scientific practitioners are dependent on the methodological principles that are specific to a paradigm. The specialist level of training that is required for any novice scientists in the field equips them to identify and solve puzzles or problems that are related to a certain paradigm. Only by completing this training can a scientist gain membership of that paradigm.

There are many reasons why scientists seem to have a greater level of immunity towards anomalies arising from their theories than historians. Kuhn explained how the experimental and methodological boundaries of their paradigm inform their investigation of nature. This makes it less likely that an anomaly will arise within the specific variables that are investigated in any experiment. Moreover, the primary motivation for a scientific experiment is that it will validate some aspect of a theory, hence the relevant variable outcomes are established prior to the experiment. Kuhn describes this as 'normal science'. Normal science is a process of 'filling-in the gaps' of a pre-established paradigm, it is 'mopping-up' the aspects of a paradigm that do not have sufficient empirical support:

Mopping-up operations are what engage most scientists throughout their careers. They constitute what I am here calling normal science. Closely examined, whether historically or in the contemporary laboratory, that enterprise seems an attempt to force nature into the preformed and relatively inflexible box that the paradigm supplies. No part of the aim of normal science is to call forth new sorts of phenomena; indeed those that will not fit in the box are often not seen at all. Nor do scientists normally aim to invent new theories, and they are often intolerant of those invented by others. Instead, normal-scientific research is directed to the articulation of those phenomena and theories that the paradigm already supplies.[43]

Both Berlin and Kuhn agree that a scientific practitioner lacking common intelligence could be regarded as a great scientist, but Kuhn emphasises that this is because normal science is opposed to critical discourse. According to Kuhn, 'it is precisely the abandonment of critical discourse that marks the transition to a science'.[44]

Paul Feyerabend takes issue with this uncritical aspect of Kuhn's description of normal science by comparing it with organised crime and lists the following similarities. Like scientific practitioners, safebreakers are solving puzzles in accordance with their expectations of phenomena. They also develop special-purpose tools for solving these puzzles. In the event of failure, the inadequacy of an individual safebreaker is blamed rather than the theory. According to Feyerabend, '[w]herever we look – the distinction we want to draw does not exist'.[45] This comparison reveals important differences between the conception of scientific practice for Kuhn and Berlin. For Kuhn, scientists are informed in the context of a paradigm that they follow uncritically when engaging in normal science. The aspects of scientific theory that Berlin regards as lacking in historical theory are merely manifestations of the uncritical structure of normal science.

Berlin also appeals to the differing uses of language for historical and scientific textbooks to demonstrate how scientists achieve a certainty that is lacking for historians. The historian weaves together logically independent concepts and events in their attempts to explain how they are causally related. For scientists, the situation is precisely the opposite; in a textbook of physics or biology 'the links between the propositions are, or should be, logically obvious'.[46] The conviction of the correctness of explanations that scientific theories yield does not depend on the rhetoric and skill of any single scientist. The explanation is considered as correct because it can be logically demonstrated. Even if the language that implied the inference (such as 'because', 'therefore', 'hence', etc.) were to be removed from all scientific textbooks, the theory should still be able to demonstrate its inner logical structure. In contrast, if this were to happen in all history textbooks, then:

> the bald juxtaposition of events or facts would at times be seen to carry no great logical force in itself, and the best cases of some of our best historians (and lawyers) would begin – to the mind as conditioned by the criteria of natural science – to seem less irresistible.[47]

For Kuhn, the function of the rational and logical structure of scientific textbooks is pedagogical. When a new scientific paradigm replaces another, then the textbooks must be re-written. According to Kuhn, textbooks 'have to be rewritten in the aftermath of each scientific revolution, and, once written, they inevitably disguise not only the role but the very existence of revolutions that produced them'.[48] Scientific textbooks establish their logical and rational rigour at the cost of their commitment to historical accuracy: '[f]rom the beginning of the scientific enterprise, the textbook presentation implies, scientists have striven for the particular objectives that are embodied in today's paradigms.'[49] *Pace* Berlin, science only *appears as* a single unified logical account of reality because of the pedagogical function of the presentation of science as a unified discipline.

The pedagogical function of scientific textbooks is to reinterpret the history of science to establish a false similarity between the research interests of their current paradigm and the interests of previous scientific theories. They claim that the puzzles of contemporary science have also been the puzzles of previous scientific paradigms, yet previous scientists did not realise this. They present their theories as offering solutions to puzzles that scientists have been tackling for centuries (if not millennia). This prevents practitioners of the current paradigm from viewing themselves as representatives of yet another paradigm that will eventually be 'overthrown' by a currently unfathomable theory. They view their own paradigm as different because it has answered the questions that have persisted throughout the history of science.

There are many examples that support Kuhn's analysis of this function that scientific textbooks serve. These kinds of argumentative strategy are not limited to scientific textbooks; they can also be found in books directed towards philosophers of science. For instance, Ernst Mayr argued that Aristotle's conception of *eidos* 'was conceptually virtually identical with the ontogenetic program of the developmental physiologist'.[50] Similarly, Ruse compares Darwin with the contemporary 'selfish gene' supporters, as 'Darwin didn't know about genes, so he could not be a "selfish gene" supporter, but he was as close to that as it is possible to be'.[51] Such statements potentially distort historical figures who have contributed to the development of science because it subjects them to the considerations of contemporary science. They implicitly offer support to the idea that science is progressive because the problems identified by the contemporary paradigm have persisted throughout the development of science, unbeknownst to the individuals who have supposedly endorsed these views. Robert Richards, in opposition to Ruse, explains how interpretations of Darwin have tended to consider Darwin through the lens of contemporary biology:

> Even good historians have been blinded by the light of modern evolutionary theory when attempting to give an account of the historical Darwin. In that brilliant glow coming from our contemporary science, those historians have constructed his doppelgänger.[52]

Strawson's interpretation of Kant is also susceptible to the criticism of constructing Kant's doppelgänger. He refused to examine the merits of transcendental idealism on its own terms, instead he approached it from the perspective of the 'scientifically minded philosopher'. Strawson's interpretation might be considered as inadvertently constructing Kant's doppelgänger. It is inadvertent because Strawson is clear from the outset that his account 'is by no means a work of historical-philosophical scholarship'.[53] Nevertheless, the impact of Strawson's interpretation has been far-reaching and long-lasting with respect to Kant scholarship.

The philosopher of history, Collingwood, suggested that it marks a maturity of thought when one realises that thought has differed at different times and places. He explains that each societal milieu is built on a number of absolute presuppositions are often regarded as lacking any historical relevance; people 'may even imagine that an absolute presupposition discovered within these limits can be more or less safely ascribed to all human beings everywhere and always'.[54] For Collingwood, the whole of philosophy should be examined from a historical perspective to appreciate the absolute presuppositions of a specific societal milieu that contributed to the development of a theory. There is little prospect of establishing universal principles that hold true for humanity in all places at all times.[55]

1.2.2 Scientific revolutions and incommensurability

Kuhn argued that the history of science comprises periods of normal science that are interrupted by scientific revolutions. Scientific revolutions occur when a scientific paradigm reaches a stage of crisis. A scientific paradigm reaches a stage of crisis when it can no longer solve puzzles in accordance with the parameters set out by that paradigm. At this point, the scientific theory becomes speculative in its attempt to solve these problematic puzzles. According to Kuhn, '[a]ll crises begin with the blurring of a paradigm and the consequent loosening of the rules for normal research'.[56] The identity of the paradigm is essentially the rules that are prescribed for its normal puzzle-solving practices. Hence, this identity is jeopardised when it engages in speculation in an attempt to find solutions for these puzzles. Combined, the unproductivity of a paradigm and the loss of identity make the paradigm vulnerable to revolution. According to Kuhn, 'the single most prevalent claim advanced by the proponents of the new paradigm is that they can solve problems that have led the old one to a crisis'.[57] The new paradigm entails the destruction of the old; Kuhn states '[i]t is hard to see how new theories could arise without these destructive changes in beliefs about nature'.[58]

A new paradigm is not accepted on the basis that it can offer a more complete explanation of the world than its predecessor; it is merely able to solve puzzles that its predecessor could not. The explanation of one scientific theory replacing another by means of revolution has been criticised on the basis that it reinforces the incommensurability between paradigms. According to Feyerabend:

> Revolutions bring about a *change* of paradigm. But following Kuhn's account of this change, or 'gestalt-switch' as he calls it, it is impossible to say that they have led to something *better*. It is impossible to say because pre- and post- revolutionary paradigms are frequently incommensurable.[59]

The incommensurability of scientific theories is similar to an aspect of Leibniz's description of monads. Monads are atoms of pure existence that cannot communicate with one another. According to Gottfried Wilhelm Leibniz, 'monads have no windows through which something can enter or leave'.[60] One difference between the kinds of incommensurability expressed by monads and paradigms is that paradigms usually successively follow one another, whereas all monads exist simultaneously. Paradigms are windowless in the sense that they are not translatable into the terms of other paradigms. Scientific textbooks are used to cover up both the revolutionary context of the origin of paradigms and their incommensurability with other paradigms. One paradigm replaces another by means of revolution; however, this is not communication between paradigms but a forceful dethroning of one paradigm for another. According to Kuhn, members of different paradigms 'do in some sense live in different worlds'.[61]

Kant considered the importance of revolution in both its philosophical and political or societal contexts. He endorsed the notion of scientific revolutions when he described the *Critique of Pure Reason* as analogous to the Copernican revolution.[62] The analogy signifies that transcendental idealism should be regarded as a break with previous philosophical accounts in the same way that the Copernican Revolution was a break with previous scientific accounts. In this sense, Kant regarded transcendental idealism as analogous with the notion of scientific 'paradigm shifts' that Kuhn described. Yet, Kant's conception of science was significantly different from Kuhn's; for Kant, science was apodictically certain and universally true. According to Michael Friedman, 'an absolutely universal human rationality realized in the fundamental constitutive principles of Newtonian science made perfectly good sense in Kant's own time, when the Newtonian conceptual framework was the only paradigm'.[63]

Kuhn's account of science as a social activity makes it possible to offer an alternative comparison of their accounts of revolution. It is helpful to compare Kant's political philosophy, which is concerned with understanding how people can respect each other's rights in a civil society, with Kuhn's conception of scientific revolutions. Kant condemns political revolutions because they are not compatible with reason, whereas Kuhn draws upon political revolutions to support their role in science. He argues that political revolutions occur in cases where political systems cease to adequately resolve problems that negatively impact on a sector of the community. In this sense, these political systems malfunction for those members of the community. According to Kuhn, '[i]n both political and scientific development the sense of malfunction that can lead to crisis is a prerequisite to revolution'.[64] He explains the vital role that political revolutions have in the evolution of political institutions.[65] The function of revolutions is to establish a change that is prohibited by the political institution itself. Kuhn regards political revolutions as an exemplar of how scientific revolutions ought to function.

In contrast, Kant opposed political revolutions for a variety of reasons. First, there can be no right to rebellion or political revolution because there is no contractual agreement between the head of state and its members; 'the relation between the people and the head of state is not contractual because there could be no neutral third party to enforce the contract'.[66] Second, revolutions are incompatible with reason as they prohibit the measure of progress. Political change must be carried out in accordance with *reformation*, not *revolution*. This allows us to work continuously on the aspects of the political institution that are malfunctioning without the cost of losing the aspects of a political system that are not malfunctioning. This is the only way to ensure that politics develops in accordance with reason. According to Kant:

> [Political change] should not be made by way of revolution, by a leap, that is, by violent overthrow of an already existing defective constitution ... But if it is attempted and carried out by gradual reform in accordance with firm principles, it can lead to continual approximation to the highest political good, perpetual peace.[67]

For transcendental idealism, the idea of political revolution is antithetical to the progress of reason. This is not because the political system replacing the current system will necessarily be worse; rather, overthrowing one system for another makes it impossible to judge whether that system is better. In this sense, Kant's argument against political revolutions is that it results in the inability to measure progress arising from such revolutions as the two political systems are incommensurable.

This difference between Kant and Kuhn is also evident from their accounts of the difference between the descriptive and prescriptive aspects of science. Kuhn's account of the structure of science blurs the distinction between these aspects of science. Hence, the practices that scientists engage in, are also the practices that they ought to engage in.[68] In contrast, Kant separates the prescriptive and descriptive aspects of human activity. An investigation of how people act is anthropological, whereas an investigation into the way people ought to act is a matter for practical reason and politics. This distinction allows Kant to separate the descriptive fact that political revolutions have occurred in a certain way, from the prescriptive claim that they ought to have occurred that way. Therefore, although Kant never considered the possibility that science was a social activity, his political philosophy helps to expose assumptions relating to Kuhn's idea of scientific revolutions. Kant pre-empted the criticism against Kuhn that revolution came at the cost of incommensurability and the inability to measure progress.

In summary, this section has explored how the understanding of the methodologies of science and history has significantly changed. For Berlin, history lacked the rigorous systematic methodology of science. For him, the fundamental difference between these methodologies was that history depended on the sagacity and skill of individual historians. In contrast, he argued that discoveries in science did not depend on any specific use of language, as they are logically demonstrable. Kuhn's analysis of the structure of science radically transformed our understanding of scientific methodology. For Kuhn, the apparent systematicity of scientific methodology came at the cost of making normal scientific practice essentially uncritical. Scientists follow the rules of a particular paradigm blindly; they are engaged in 'mopping-up operations' that neither promote the emergence of unexpected results nor make sense of such results if they arise. Feyerabend emphasised the uncritical aspect of Kuhn's account of science by comparing the activities of practitioners engaged in normal science to practitioners of organised crime. The incommensurability between different scientific theories is a consequence of Kuhn's argument that transitions in science are achieved by means of revolution. Different scientific paradigms at any one time are untranslatable to one another; a scientific practitioner can only move between paradigms by a process analogous to a conversion experience.[69] Scientific theories have pedagogical mechanisms for covering-up the occurrence of these revolutions through textbooks that present the history of science as culminating in the problems that can be solved by the current paradigm. In relation to Berlin's account, this reveals how science distorts its historical emergence by suggesting that previous sciences were unconsciously engaged in the problems specific to contemporary science. Kant's criticism of political revolutions reveals another issue with Kuhn's appeal to political revolutions as justification for scientific revolutions. Kant opposed political revolutions because they resulted in incommensurability and denied knowledge of any possible societal progress. Hence, we can draw from Kant's political philosophy to object to Kuhn's account of scientific revolutions on the basis that his account stands in opposition to the rationality and progress of science.

1.3 The context of research programmes and the Lenoir thesis

Lakatos's account of research programmes developed from the shortcomings of Kuhn's philosophy. Lakatos took issue with Kuhn's account of anomalies. For Kuhn, scientists approach anomalies blindly; however, Lakatos argued that scientists anticipate the impact of potential anomalies and direct their research accordingly. I outline Lenoir's appeal to Lakatos's conception of research programmes in relation to his account of Kant's influence on the development of nineteenth-century German biology. I examine criticisms of Lenoir's account and argue that these should be understood as criticisms of Lenoir's appeal to Lakatos, rather than of Kant's influence on the development of biology more generally. This lays the foundation for an alternative account of influence developed in the next section.

1.3.1 Lakatos's conception of research programmes and the Lenoir thesis

Lakatos's conception of research programmes exposes important limitations of Kuhn's account of the structure of science and scientific practice. In addition, Lenoir adopts Lakatos's conception of research programmes as a guiding thread for understanding the historical influence of Kant on the development of nineteenth-century German biology. Importantly, Lakatos builds on the foundations of Kuhn to develop a methodology of science that emphasises the rationality of science. His account of research programmes must be discussed before examining its deployment within Lenoir's thesis.

Lakatos criticises the shortcomings of Kuhn's account of the structure of science because it had not sufficiently explained how science deals with the potential anomalies that arise during the activities of normal science. Kuhn had merely explained that anomalies were not an expected outcome of normal scientific activity, they emerged because of the difficulties regarding the 'paradigm-nature fit'. According to Kuhn, 'if an anomaly is to evoke crisis, it must usually be more than just an anomaly. There are always difficulties somewhere in the paradigm-nature fit; most of them are set right sooner or later'.[70] Science must be carried out in accordance with a paradigm, and paradigms are opposed to the discovery of unexpected evidence. In short, normal science opposes the emergence of novelty: 'novelty emerges only with difficulty, manifested by resistance, against a background provided by expectation'.[71] The occurrence of novelty suggests that the expectations of the scientific community are wrong and the best instruments that we have for understanding the world are not fit for purpose. According to Kuhn, '[u]nanticipated novelty, the new discovery, can emerge only to the extent his anticipations about nature and his instruments prove wrong'.[72] Lakatos criticised Kuhn's account of anomalies because he did not consider how scientific method could be internally structured to target certain anomalies. For Lakatos, scientific methodology directs research towards anomalies that are more likely to be resolved and away from anomalies that would be detrimental to the scientific paradigm or research programme more generally. In other words, science has a mechanism for anticipating the potential detriment of an anomaly and can direct research towards anomalies that are less detrimental.

Lakatos compares anomalies to an ocean. Without intervention, the ocean will erode a shoreline in much the same way that anomalies will erode a paradigm (or what Lakatos calls a 'research programme'). The erosion of the shoreline is prevented by creating coastal defences and redirecting this eroding force. Analogously, Lakatos argues that scientific theories have two distinct heuristic functions: positive heuristics and negative heuristics. The negative heuristics of a research programme relate to the elements of a theory that scientific practitioners must subscribe to as a condition of identifying themselves as belonging to that research programme. If these principles are abandoned or refuted, then the research programme collapses. Scientific research is directed away from any potential research that will bring these fundamental principles (the negative heuristic or hardcore principles) of the theory into question; instead, practitioners direct their research towards auxiliary hypotheses. Auxiliary hypotheses can be refuted without implicating the hardcore elements themselves. The refuted auxiliary hypothesis can then be replaced with an alternative auxiliary hypothesis. In effect, they form a protective belt around the central hardcore principles. This is a defensive strategy to make sure that the possible anomalies that could emerge from scientific research are not fatal for the research programme. According to Lakatos:

> [I]t should not be thought that yet unexplained anomalies – 'puzzles' as Kuhn might call them – are taken in random order, and the protective belt be built up in an eclectic fashion, without any preconceived order. The order is usually decided in the theoretician's cabinet, independently of the *known* anomalies. Few theoretical scientists engaged in a research programme pay undue attention to 'refutations'. They have long-term research policies which anticipate these refutations. This research policy, or order of research, is set out … in the *positive heuristic* of the research programme. The negative heuristic specifies the 'hard core' of the programme which is 'irrefutable' by the methodological decisions of its protagonists; the positive heuristic consists of a partially articulated set of suggestions or hints on how to change, develop the 'refutable variants' of the research programme, how to modify, sophisticate, the 'refutable' protective belt.[73]

For Lakatos, the drive towards the self-preservation of a research programme that can anticipate the eroding force of potential refutations marks the development of a mature science. In contrast, an immature science is one that proceeds by a mere patchwork of trial and error. Mature science anticipates anomalies by employing a mutual interplay of rationalism and dogmatism. He argues that only by considering science as a battleground of competing research programmes rather than isolated theories can we understand 'the rationality of a certain amount of dogmatism'.[74] This dogmatism consists in weighing-up the importance of different aspects of a research programme and deciding whether these features belong to the positive or negative heuristics of that research programme. The hardcore elements or negative heuristics are then protected from potential anomalies by directing research towards auxiliary hypotheses.

The idea that this is a rational part of scientific activity is questionable. The pragmatic reasons for protecting the hardcore elements of a research programme are self-evident as the refutation of these principles results in the collapse of the research programme, but this does not mean that such protection is rational. The emerging picture of science is that the most fundamental aspects of a research programme are protected from empirical and rational criticism because of the potentially fatal implications for the research programme if these principles are shown to be wrong. Instead, enquiry is directed towards the expendable auxiliary hypotheses that can be replaced. In the context of the development of science, such protective measures might be permissible for the development of scientific research programmes. This is a side issue, as here the primary concern is not with the *rationality* or *arationality* of Lakatos's account in the context of the philosophy of science, but with the deployment of his theory of research programmes to understand Kant's influence on the development of biology.

Lenoir appealed to Lakatos's account to explain Kant's influence on the development of nineteenth-century German biology. According to Lenoir, developmental morphology, cell theory and functional morphology were all indebted to Kant's *Critique of the Power of Judgment*. Kant forms the hardcore element or negative heuristic of the scientific research programme that he terms teleomechanism. Lenoir argues that Kant's conception of the morphotype was part of the negative heuristic of the teleomechanist research programme. The hardcore elements of a scientific research programme 'can never be the object of empirical refutation, and ... cannot be abandoned without repudiation of the program'.[75] The teleomechanists were unified by 'an expression of commitment to the holistic conception at the heart of the teleomechanist research tradition'.[76]

1.3.2 Criticisms of the Lenoir thesis

Lenoir's thesis has generated much controversy. The problem is that Kant's discussion of teleology in the third *Critique* is not consistent with its application in biology. His account has been the focus of criticism because it excludes other figures that were also influential on the development of German biology. Moreover, others have argued that the underlying incompatibility between transcendental idealism and naturalism threatens to invalidate the Lenoir thesis. I argue that both these criticisms expose problems for the account of influence that Lenoir adopts, but do not necessitate the conclusion that Kant did not influence the development of biology. Rather, the conception of influence permitted by the notion of research programmes is too restrictive to offer an appropriate understanding of Kant's influence on the development of biology.

Lynn Nyhart argues that Lenoir's account does not consider the variety of influences beyond Kant that contributed to the development of German biology. She argues that: '[m]ost early nineteenth-century writers on form confound categorisation schemes based on rigid philosophical distinctions; they appropriated the language of Kant, of Schelling, and of Cuvier in different places … [M]any had little apparent trouble drawing selectively from … philosophically opposed systems.'[77] The idea that biologists drew selectively from opposed systems is problematic for Lenoir's indebtedness to Lakatos. Lakatos understood that many developments in science have been achieved by means of 'grafting' a new programme onto an inconsistent older programme. He argues the two programmes cannot persist in this symbiotic relationship. According to Lakatos, '[a]s the young grafted programme strengthens, the peaceful co-existence comes to an end, symbiosis becomes competitive and the champions of the new programme try to replace the old programme altogether'.[78] Chaos grows within the research programme until it is purged of these inconsistencies. 'Grafted' programmes serve an important function for emerging research programmes as they can exploit their heuristic power over older programmes. However, these inconsistencies cannot reside permanently in a research programme. 'The reason is simple. If science aims at truth, it must aim at consistency; if it resigns consistency, it resigns truth.'[79]

It follows from Nyhart's account that the influences on the development of German biology could not be easily understood as culminating in a single, consistent research programme. Lenoir's appeal to research programmes as a way of understanding the development of German biology creates an obstacle for appreciating the various figures that these biologists were influenced by. The norms that Lakatos identifies as governing scientific research cannot adequately explain how various influences that are philosophically opposed to one another could have contributed to the development of German biology.

The unsuitability of Lakatos's conception of research programmes for understanding Kant's influence on the development
of biology relates to the broader incompatibility between the principles of transcendental idealism and naturalism. Both Zammito
and Richards have criticised the Lenoir thesis because it can only
establish the connection between Kant and biology by overlooking
significant differences between these accounts. They focus on Kant's
influence on Blumenbach to expose an underlying insurmountable
difference. In contrast, Lenoir emphasises the mutual support that
these philosophers believed their accounts had for one another. In
a letter to Blumenbach, Kant explicitly noted his indebtedness to
Blumenbach's essay *On the Formative Impulse*. He recognised the
similarity between Blumenbach's conception of the formative drive
(*Bildungstrieb*) and his attempt to unify teleological and mechanistic explanations of organised nature; 'factual confirmation is
exactly what this union of the two principles needs'.[80] Kant was
inspired by Blumenbach's view that the emergence of entities that
organise their parts and their whole through a special drive could
not be explained by appealing to merely mechanical laws. According to Kant:

> [H]e rightly declares it to be contrary to reason that raw matter should
> originally have formed itself in accordance with mechanical laws, that
> life should have arisen from the nature of the lifeless, and that matter
> should have been able to assemble itself into the form of a self-preserv-
> ing purposiveness by itself.[81]

There is an important difference between the projects of Kant
and Blumenbach. Kant explained how our judgements of organisms
could not reveal real teleological causes in the objects themselves.
Instead, we are limited to conceiving of teleological entities – that is,
entities that require us to judge them as end directed – as a product
of the judgement itself. This distinction means that we are not justified to assert that our judgements pertaining to biological principles
refer to properties of the entity. We judge the entity *as if* the features of the judgement applied to features that are external to that
judgement. Therefore, Kant is not offering support to the biological
sciences, as biologists do not consider the properties specific to
organic nature as dependent on our faculty of judgement. Blumenbach proposed that each organism possessed an internal formative
force or *Bildungstrieb* that was responsible for its organisation:

Bildungstrieb was thus not a Kantian 'as if' cause but a real teleological cause (i.e. one acting for ends), which, albeit, was known only through the ends it achieved ... Blumenbach clearly spied the Creator unabashedly pulling the strings, a perception no scientific theory in the Kantian mold would legitimate.[82]

According to Kant, judging biological entities as if they are governed by teleological principles discloses nothing more than the act of judging them to possess this property. The transcendental idealist account of the organism considers our judgement of organisms to be a mirror exposing our own capacity to judge nature teleologically. Kant denied biology from the status of science because our ability to identify biological entities depends on our power of judgement. In contrast, Blumenbach regarded our ability to perceive the teleology of organisms as providing a window from which we see the intentions of the supersensible Creator; this was beyond the remit of knowledge according to Kant's account.

Richards and Zammito are correct to question the consistency of Lenoir's thesis insofar as he does not sufficiently address the implications of Kant's conception of biological phenomena as the product of regulative judgements. The *apparent* similarity between Kant and Blumenbach hides a deeper divergence between their philosophies. However, Lenoir is also aware of this as he recognised that the scientists of the teleomechanist research programme transformed Kant's regulative principles into constitutive ones. Lenoir notes, 'Kielmeyer as a biologist found it difficult to remain consistent with this regulative use of the principle ... [H]e had overstepped the valid limits of the concept of teleology as Kant formulated it'.[83] Lenoir did not consider these biological tendencies to go beyond Kant's regulative principles as destructive for the research programme. He clarifies the intentions of his project: 'I do not claim that German biologists discovered the programme of research in Kant's writing which they set out to realize in practice ... [Kant] set forth a clear synthesis of the principal elements of an emerging consensus among biologists.'[84]

Kant's distinction between regulative and constitutive principles is an obstacle to understanding how Kant could form the hardcore elements of the teleomechanist research programme. It is unclear in what sense Kant could be regarded as the hardcore element of the research programme given the tendency of its members to apply

Kant's regulative principles constitutively. Kant's third *Critique* was attempting to resolve issues that were specific to transcendental idealism. The subsequent biological concerns with the organism were not engaging with these issues. According to Zammito, biology is founded on the principles of naturalism 'and Kant simply cannot be refashioned into a naturalist'.[85]

In summary, Lakatos's conception of research programmes is not appropriate for understanding Kant's influence on biology. Lenoir did not consider the other possible sources of influence on German biology (as Nyhart argues) because he developed his account of research programmes from Lakatos. He also overlooks the fundamental incompatibilities between transcendental idealism and biology (as Zammito argues). In the next section, I develop the foundations of an alternative account of influence that reveals how these criticisms do not invalidate claims regarding Kant's influence on biology, rather they only challenge Kant's influence on biology as understood through the conception of research programmes.

Lenoir's thesis is a useful resource for understanding the potential difficulties that can emerge when investigating Kant's influence on biology. The method that Lenoir adopts to examine the influence of Kant on the development of biology in nineteenth-century Germany is very different from my own analysis of Kant's influence on the development of biology in the British Isles. My analysis examines the significance of aspects of Kant's philosophy for development of science, while also exploring how these ideas were transformed from their original Kantian formulation in ways that were appropriate for this emerging scientific method. This analysis is indebted to those previous accounts and the controversies and criticisms raised in relation to them, which helps us to understand what is required from an account of influence that aims to present Kant's philosophy in a manner that is both philosophically relevant and historically accurate.

1.4 Expanding the scope of influence

The problem specific to the previous accounts of influence discussed is that they have implicitly presupposed that the source of an influence must be compatible with its subsequent effects. The requirement of compatibilism places an unrealistic demand on what counts as an influence that significantly limits the scope of potential sources of influence. I propose an alternative account influence that focuses on how ideas can be influential even if they are intentionally transformed or unintentionally misunderstood to be more suitable for the conceptual demands arising from the subsequent theories that they are applied to.

I draw upon several philosophers to offer a variety of argumentative strategies that support a broader account of influence. First, I examine Bloom's account of poetic influence, which argues that intentional misunderstanding plays an essential role in the development of poetry. Bloom explains how poetic influence does not presuppose the condition of similitude as poets intentionally misrepresent one another through the act of misprisions to establish their own poetic identity and novelty. Second, Feyerabend makes a similar claim through his distinction of the context of the discovery of a scientific theory from the context of its justification. For Feyerabend, this forms the basis of his argument that the development of science is 'against method'. In contrast, I argue that this account helps to establish a method for understanding how influence can arise out of creative misunderstandings. Third, Potochnik considers how many aspects of scientific development are based on idealisations. The aim of these idealisations is not furthering truth, but only our understanding of an aspect of science. This account offers support to my argument that potential sources of influence or inspiration for scientific theories do not necessarily offer accurate representation. Finally, I examine Collingwood's account of metaphysics as absolute presuppositions. I consider how these alternative accounts of influence could offer support to Zammito's claim that the development of biology was caused by a misunderstanding of Kant's philosophy.

1.4.1 Bloom: influence in poetry

The general definition of 'influence' is the capacity to have an effect on the character, development or behaviour of something or someone. According to Bloom, the etymological connotations of influence have been lost, and 'to be influenced meant to receive a divine power; to receive an ethereal fluid ... from the stars'.[86] The reason why Bloom focuses on the etymology of influence is to demonstrate that, historically, it lacks clarity and there is no agreement for a precise meaning of the term. The meaning of influence is something that seems to be clear from a general perspective, but when disagreements arise about appropriate sources of influence, it is far from clear how these disputes can be resolved.

Bloom's account of the role of influence within poetry opposes many of the assumptions that are implicit in the scientific understanding of influence discussed in the previous section. Bloom explains how it is commonplace for poets to intentionally misinterpret or creatively correct their predecessors to establish their own poetic identity. A poet is not limited by the norms of a research programme or by scientific training, which oppose the emergence of anomalies. In poetry, creative freedom of interpretation is a means of establishing their own legacy. For Bloom, these creative acts are not rational strategies on the part of the poet, rather they are a response to the poet's anxiety about being forgotten. According to Bloom:

> Poetic Influence – when it involves two strong, authentic poets – always proceeds by a misreading of the prior poet, an act of creative correction that is actually and necessarily a misinterpretation. The history of fruitful poetic influence, which is to say the main tradition of Western poetry since the Renaissance, is a history of anxiety and self-saving caricature, of distortion, of perverse, wilful revisionism, without which modern poetry as such could not exist.[87]

Bloom argues that the history of poetry cannot be separated from this account of poetic influence. The history of poetry can be understood only by identifying how poets have misunderstood one another. He explains his examination of the history of the trajectory of poetic influence by appealing to Lucretius's notion of *clinamen*. This refers to the unpredictable and spontaneous free movement (or swerve) of atoms in a void that was central to the philosophy of Democritus. He argued that the cosmos consisted of atoms falling in an infinite void and the production of objects must have been the result of one atom spontaneously swerving into another. Poets also manifest these spontaneous 'swerves' through acts of intentional misrepresentation (creative correction or misprision) that are the cause of them diverging from their precursors. According to Bloom, 'the true history of modern poetry would be the accurate recording of these revisionary swerves'.[88]

Bloom's discussion of influence reveals how alternative accounts of influence can significantly differ from the scientific accounts of influence discussed in previous sections. The role of influence that underpinned Kuhn's and Lakatos's accounts of scientific method focused on training and specialisation, which imposes strict limits on possible sources of influence. Discussion between alternative paradigms is limited because of difficulties surrounding commensurability, and experiments tend to oppose the emergence of novelty. Bloom shows there are alternative ways that influence can manifest. In poetry, established poets will influence emerging poets, but emerging poets will creatively misrepresent these poets through acts of misprision to establish their own identity.

1.4.2 Feyerabend: science as a creative process

The philosopher of science, Feyerabend, has also argued that the development of scientific theories is essentially a creative process, which originates by irrational or *arational* means. He uses the distinction between the context of discovery and the context of justification: '*[d]iscovery* may be irrational and need not follow any recognized method. *Justification*, on the other hand ... starts only *after* the discoveries have been made, and it proceeds in an orderly way.'[89] This distinction, originally formulated by Hans Reichenbach, was intended to separate the contextual factors that determine the validity of a theory from the factors that are specific to its conception. The former relates to the context of justification whereas the latter relates to the context of discovery. Reichenbach explains the difference between these contexts with the example of being faced with a mathematical problem:

> If any solution is presented to us, we may decide unambiguously and with the use of deductive operations alone whether or not it is correct. The way in which we find the solution, however, remains to a great extent in the unexplored darkness of productive thought and may be influenced by aesthetic considerations, or a 'feeling of geometrical harmony.'[90]

For Reichenbach, the context of discovery relates to those aspects of a theory that cannot be subject to critical examination. Thus, Reichenbach distinguishes between the contexts of justification and discovery on the basis that the former refers to the critical aspects of a theory whereas the latter is descriptive. Karl Popper compared this aspect of Reichenbach's account with Kant's distinction between questions of fact and questions of justification. Aspects of a theory belonging to the context of discovery can only be explained factually or descriptively. Popper rejects the relevance of these aspects of a scientific theory for the analysis of scientific knowledge. According to Popper, '[t]he question of how it happens that a new idea occurs to man ... may be of great interest to empirical psychology; but it is irrelevant to the logical analysis of scientific knowledge'.[91] Reichenbach does not completely reject the importance of the context of discovery for scientific knowledge, as many aspects of a scientific theory fall within this domain. For instance, he argues that conventional aspects of a theory, such as units of measurement, belong to the context of discovery and these features constitute 'an integral part of the critical task of epistemology'.[92]

The difference between Feyerabend's and Reichenbach's distinctions of the contexts of discovery and justification is that Feyerabend argues these are temporally distinct moments in the development of a theory. The context of the discovery of a scientific theory and its subsequent justification are different stages with different conceptual requirements. These conceptual requirements are not merely different from one another, they are in *conflict* with one another. The discovery or invention of a new theory is not restricted to the methodological rules by which science is subsequently justified.

The idea that the discovery of a scientific theory is not always a rational process allows us to expand the scope of potential sources of influence. Influence is limited only by the creative imagination of the individual, not the compatibility with that scientific theory. This offers support to the idea that aspects of Kant's philosophy could have been influential for the development of biology despite its incompatibility with naturalism.

1.4.3 Potochnik: idealisation and science

The idea that scientific models and theories are entirely naturalistic is also the subject of dispute. Although Potochnik does not align her position to either realism or anti-realism, she argues that the advancements in many scientific models and theories do not correspond with how closely those models and theories represent

nature. Her account focuses on scientific practice as a psychological activity directed towards understanding rather than truth:
'whether a proposition is true in no way depends on the psychology
of one who entertains or believes that proposition.'[93] Replacing the
pursuit of truth with the pursuit of understanding reorientates us
towards science as a cognitive and epistemic practice. These features hold the key to explaining why science is full of '*rampant and
unchecked*' idealisations.[94]

The irreducible relationship that scientific practice has with our
cognitive understanding of the world entails that science 'depends
in part on the psychological characteristics of those who seek to
understand'.[95] These psychological characteristics range from
social values such as decisions by funding bodies, to our fundamental human limitations such as our relatively short life spans.
Through idealisations, we can understand very complex phenomena that would otherwise remain incomprehensible. Unveiling the
role of idealisations within scientific practice does not undermine
scientific achievements, but rather reveals that the aim of science is the epistemic goal of understanding, rather than seeking
to uncover metaphysical truth. For idealisations to be successful
for science, they must pass what Potochnik terms the 'standards of
success',[96] which refers to how idealisations isolate specific causal
patterns that are not identifiable under a more accurate portrayal
of the world. A paradigm example of this is the focus on genetic
explanations in contemporary philosophy of biology, rather than
environmental ones. The reason for this predominance might well
relate to societal values and policies driving such research. According to Potochnik, '[w]hat is problematic is attributing the focus on
genes to how the world really is, to the relative causal importance
of genes over environment, rather than to simply what human
observers have most often sought to understand'.[97] This in no
way denies that genetic influences identify real casual patterns in
nature; Potochnik's concern is that we place greater metaphysical
importance on genetic influences because of the greater amount of
attention it has received.

The idea that idealisations reflect the aims of scientific practitioners, rather than revealing fundamental metaphysical truths
about nature, is important for two reasons. First, it reveals that it
is not necessary to have direct representation between our experience of nature and scientific theories and models. Science does not

hold to naturalism as the governing principle for its development. Potochnik discusses this in terms of the commitment to realism or anti-realism. She argues that her emphasis on idealisations constrains the possible versions of realism or anti-realism that will be appropriate for philosophy of science; however, it does not rule out either realism or anti-realism.

Second, science comprises a large number of idealisations that do not come together to form a unified conception of science. There are no conditions for idealisations in science beyond their ability to isolate causal patterns with precision. The same phenomenon can be part of different idealisations that isolate different causal patterns. As idealisations are not of direct metaphysical significance, it is not possible to impose a metaphysical system on idealisations as a way of ordering them.[98]

1.4.4 Collingwood: metaphysics as absolute presuppositions

The philosopher of ideas, Collingwood, raises similar concerns against the superiority that is generally ascribed to metaphysics within philosophy. In a similar fashion to Kuhn's revelation that history and practice were crucial for understanding scientific method, Collingwood argued that metaphysics was also essentially historical. According to Collingwood:

> Metaphysics, aware of itself as a historical science, will abolish in one clean sweep not only the idea of 'schools' but also 'doctrines'. It will realize that what are described as A's 'metaphysical doctrines' are nothing more than the result of A's attempt to discover what absolute presuppositions are made by scientists in his own time.[99]

Absolute presuppositions relate to the way that we can manipulate and convert the contents of experience to inform science and civilisation. Absolute presuppositions shape the way we comprehend experience; Collingwood describes them as 'the yard-stick by which 'experience' is judged'.[100] Absolute presuppositions have varied between people at different times and places. Collingwood emphasises this by considering a group that believes in magic: '[a]s long as you measure in feet and inches, everything you measure has dimensions composed of those units. As long as you believe in a world of magic, that is the kind of world in which you live.'[101] Collingwood is describing a version of what Kuhn later terms the incommensurability between scientific theories.[102] For those who believe in magic, there is no possible experience that would offer sufficient evidence against their belief. Of course, Collingwood does not deny that their beliefs might change, but this would more likely be due to changes in beliefs in powerful members of their own community or other respected communities who previously held similar absolute presuppositions that have now changed. Again, this offers support to the idea that developments in science are not always produced in accordance with rational processes. Absolute presuppositions are not derived from experience, rather they are the way in which our minds understand experience.

One essential divergence between absolute presuppositions is whether a society is polytheistic or monotheistic. Collingwood argues that Ancient Greek society was committed to a form of polytheism,[103] which had a tendency to explain events poetically. Their explanations of events were often in terms of disputes between quasi-human manifestations of gods. For instance, Aphrodite was 'a superhuman will, brought about by the various events which together made up her realm, namely the events connected with sexual reproduction'.[104] For modern science to develop, it was necessary a move away from explanations of natural events in terms of interactions between gods. Instead, it required us to seek explanations of natural events in terms of universal natural and mathematical laws. According to Collingwood, '[t]he attempt to replace a polymorphic by a monomorphic natural science was logically bound up with the attempt to replace a polytheistic by a monotheistic religion'.[105] Collingwood's account of science as essentially monomorphic reveals the underlying importance of the unity of science for his account. This is in contrast with Potochnik's critical stance against the unity of science because it is not possible to find evidence to show that certain idealisations are more fundamental than others. Collingwood argues that the possibility of science requires that we must first possess the absolute presupposition that the world is the product of a monotheistic God.[106]

1.4.5 The emerging conception of influence

When we consider the accounts from Bloom, Feyerabend, Potochnik and Collingwood in combination, they allow us to expand the conception of influence and recognise how developments in science can be inspired by moments of creativity that do not follow strict rules. Bloom revealed that it is commonplace for poets to establish their own identity through acts of misprision; in other words, intentionally misunderstanding or creatively correcting their predecessors. Feyerabend emphasised the creativity inherent with the development of scientific theories through his distinction between the contexts of discovery and justification. Moreover, philosophers of science are not necessarily committed to biological naturalism as a metaphysical position. Potochnik revealed that scientific practice is rampant with idealisations. Idealisations allow scientists to identify causal patterns that would otherwise be beyond the remit of observation, which often increase our scientific understanding of particular causal processes, but this does not necessarily correlate to truth. Scientific theories and models often diverge from a strict correspondence with nature in order to attain greater scientific understanding. We must use idealisations because of biological limitations that impose severe limits on our comprehension of nature, but idealisations also reflect the sociopolitical factors and values of science within a specific social milieu. Therefore, the notion that these idealisations provide us with a metaphysical account of reality aimed at truth is deeply misguided. Collingwood was also critical of the elevation of metaphysics over other aspects of philosophical investigation. He argued that metaphysical principles should be understood as historical absolute presuppositions that have differed between societies existing at different times. Thus, science is a fundamentally social activity that cannot be separated from various contextual factors specific to the development of scientific theories.

Combined, these accounts provide the toolkit required to develop a more accurate explanation of Kant's influence on biology. Recall that for Lenoir, Kant constituted the hardcore principles of the research programme of teleomechanism. It followed that research is directed away from the hardcore principles towards auxiliary hypotheses. Lakatos described these auxiliary hypotheses as forming a protective belt around the hardcore principles of the research programme. The possible inconsistences of the hardcore principles within the scientific research programme are protected from exposure by the mechanisms that direct research towards auxiliary hypotheses. According to Feyerabend:

It is clear that allegiance to new ideas will have to be brought about by means other than arguments. It will have to be brought about *by irrational means* ... We need these irrational means in order to uphold what is nothing but a blind faith until we have found the auxiliary sciences, the facts, the arguments that turn faith into sound 'knowledge'.[107]

The reason that Feyerabend argued that his account of the philosophy of science was '*against method*' is also the basis for incorporating his account into the foundations of an alternative understanding of influence. He argues that the source of an influence can be irrational, but this also makes it possible to consider Kant's influence on the development of biology from a different perspective.[108] In relation to Lenoir's account, it provides the basis for understanding Kant's influence on nineteenth-century German biology without presupposing Kant's compatibility with biology. According to Kant's critical philosophy, the possibility of experiencing organisms is dependent on our capacity to judge entities in accordance with final causes. In contrast, biologists tend to regard the features of organisms that Kant relegates to judgement as capacities of the organism independent of judgement. This difference led Zammito to conclude that transcendental idealism is irrelevant and useless for contemporary philosophy of biology. According to Zammito, '[i]f biology must conceptualize self-organization as actual in the world, Kant's regulative/constitutive distinction is pointless in practice and the (naturalist) philosophy of biology has urgent work to undertake for which Kant turns out not to be very helpful'.[109] Approaching the incompatibility between transcendental idealism and naturalism from a historical perspective allows us to appreciate the importance of Zammito's criticism, but also simultaneously recognise that many biologists have appealed to aspects of Kant's philosophy that are not compatible with naturalism. Kant's influence on biology was only possible because biologists creatively applied principles from Kant's philosophy as support for their research and theories. Potochnik argues that scientific models and theories regularly rely on idealisations that deviate from the position of scientific naturalism because the aim of science is not truth but understanding. These models and theories guide science despite the lack of direct correspondence with the world. In this sense, the problem raised for philosophers of biology by appealing to Kant's philosophy in part corresponds to the more general difficulty that idealisations are often used in science.

It is inconsistent for Zammito to argue that Kant was of minimal influence for the development of biology because of the incompatibility between transcendental idealism and naturalism. He even concedes that Kant was an essential influence on the development of biology; '[o]nly by misunderstanding Kant did biology as a special science emerge at the close of the eighteenth century'.[110] For Zammito, Kant could not be significant for the development of biology because these influences misunderstand fundamental aspects of Kant's philosophy. While Kant was misunderstood by biologists, this is not sufficient to demonstrate Kant's lack of importance for the development of biology. As Zammito remarks, this misunderstanding was essential for the development of biology. Hence, the tension within Zammito's account is that Kant was both causally responsible for the development of biology, and the original formulation of Kant's philosophy has little to offer biology. One way to resolve this tension is to separate influence from the condition of similarity. My account is not opposed to Zammito's claim that there is an incompatibility between naturalism and transcendental idealism. However, under my interpretation I suggest that we should pay special attention to those aspects of naturalist philosophies that have appealed to transcendental idealism in their development, as it is important to assess how they have justified these principles within the remit of naturalism. I suspect that they will struggle to justify these principles because of the incompatibility between transcendental idealism and naturalism. This does not demonstrate an absence of influence, but rather exposes the need to re-examine these aspects of their theories that appealed to Kant's critical philosophy and understand why these issues could not be resolved through naturalism alone.

Conclusion

The account of influence developed in this chapter demonstrates how transcendental idealism could have influenced the development of biology despite its incompatibility with naturalism. Some biologists misunderstood Kant and appealed to his philosophy as supporting their account of biology. This misunderstanding was important for the development of biology because it allowed biologists to resolve problems or stimulate research based on principles that went beyond the limits of what could be justified in accordance with naturalism. This serves as justification to re-address the Kantian legacy of biology, for biologists need to address the aspects of their theories that were made possible by this misunderstanding.

2 • Kant's Response to Hume and the Status of Laws in Contemporary Philosophy of Science

Introduction

Kant misunderstood Hume's distinction between relations of ideas and matters of fact. In Kant's view, Hume regarded relations of ideas as logical truths that required no justification from experience. Yet Hume in fact argued that relations of ideas were psychological truths that were derived from experience. Kant's misunderstanding hid a deep similarity between Hume's relations of ideas and his conception of the synthetic *a priori*. Kant thought transcendental idealism was responding to the shortcomings of Hume's empiricism. For Kant, transcendental idealism enabled the move beyond Humean scepticism regarding the impossibility of providing any rational justification for our belief in causality. Kant famously asserted that his remembrance (*Erinnerung*)[1] of Hume awoke him from his dogmatic slumber.[2] Transcendental idealism was not merely a response to Hume; Kant regarded it as a continuation of the project started by his predecessor. The *Critique of Pure Reason* was 'the *elaboration* of the Humean problem in its greatest possible amplification'.[3] In the first section I argue that the importance of Hume's philosophy for Kant is an example of how influence can occur despite arising from misunderstanding. Examining Kant's misunderstanding helps to uncover both the differences and similarities between their philosophies, moreover it helps to elucidate some of the fundamental principles of Kant's philosophy. It helps us to understand that the transition from Hume's naturalist stance to Kant's transcendental idealist stance is not as counter-intuitive as it might first seem.

Kant also identified some important aspects of Hume's philosophy that are overlooked by contemporary interpretations of Hume's philosophy such as the sceptical realist interpretation. In the second section I contrast Kant's interpretation with the sceptical realist interpretation, which argues that although Hume was an epistemological anti-realist regarding necessary causal laws, he was nonetheless committed to an ontological realist position about the existence of such laws. Hume dogmatically believed in the existence of laws of nature despite conceding that this belief cannot be rationally justified. Kant took Hume in a different direction. He thought Hume was entirely correct in his assessment of the impossibility of providing a rational justification for causality, if we understand causality as a necessary connection between things in themselves. Yet, Kant aimed to re-establish the importance of reason by separating objects in themselves from experiences of objects and arguing that causal necessity could only relate to experience. Approaching Kant's philosophy as a response to Hume helps to navigate through some of the core debates in Kant scholarship such as the distinction between two-world and two-aspect accounts.

The idea that causal necessity and causal laws can only relate to experience has significant ramifications for contemporary debates regarding the metaphysics of the laws of nature. In the third section I reformulate Roy Bhaskar's critical realism and Nancy Cartwright's nomological pluralism as a Kantian mathematical antinomy to reveal how both their arguments depend on conditions of experience applying to metaphysical arguments beyond experience. Kant's philosophy reveals how Bhaskar's rationalist arguments in support of the existence of the laws of nature and Cartwright's empiricist arguments against the existence of laws of nature both assume that it is possible to derive knowledge about the metaphysical status of reality from experience. Adopting a transcendental idealist perspective makes it possible to combine Cartwright's argument, that experience opposes the existence of universal laws of nature, with Bhaskar's argument that laws of nature are necessary for the possibility of science. From a Kantian perspective, this necessity does not relate to the ontological status of laws of nature, but only to our need to judge nature as if universal laws of nature govern it.

2.1 Kant's interpretation of Hume

When examining Hume's influence on Kant, many scholars tend to focus on Kant's treatment of Humean causality. Karin De Boer provides an enlightening overview of the positions that various scholars have adopted when explaining Kant's response to Hume's account of causality.[4] In contrast, I consider Kant's broader interpretation of Hume's distinction between matters of facts and relations of ideas. Kant misunderstood Hume's distinction between matters of facts and relations of ideas. He interpreted Hume as making a logical distinction between analytic relations of ideas on the one side and synthetic matters of fact on the other. Hume justified relations of ideas as psychological certainties, rather than logical ones. Relations of ideas require empirical justification, but this kind of justification is fundamentally different from the justification of causal regularity as a form of what he terms 'habit' or 'custom'. This misunderstanding resulted in Kant overlooking the similarity between his account of synthetic *a priori* propositions and Hume's relations of ideas, as both were directed towards establishing necessity from experience.

2.1.1 The sources of Kant's interpretation of Hume

Kant's interpretation of Hume is best approached by first considering the sources from which Kant is known to have derived his interpretation. Kant was most familiar with Hume's *Enquiry Concerning Human Understanding*, which was translated into German in 1755. A complete translation of Hume's *Treatise of Human Nature* (translated by Ludwig Heinrich Jacob) was not available until 1792 (although the last chapter of book one of the *Treatise* was translated for the *Königsberger Zeitung* in July 1771).[5] This means that, to the best of our knowledge, Kant could not have been aware of the complete content of this work until after the publication of the third *Critique*. Kant also acquired a copy of Schreiter's translation of Hume's *Dialogues Concerning Natural Religion* in 1881.[6]

Many consider the inaccessibility of the *Treatise* as a contributing factor to Kant's inaccurate representation of Hume's account of causality. For instance, Lewis White Beck argued that Kant drew heavily from John Beattie's commentary on Hume's *Treatise*.[7] This commentary possibly misled Kant to believe that the *Treatise* contained a more substantial treatment of causality than it did. He suggests that if Beattie had not misrepresented the *Treatise*, then Kant might not have been motivated to develop his account of transcendental idealism. Hence, Beck describes this as 'a fortunate historical error'.[8] Beattie's explanation, according to Beck:

> may have misled Kant into thinking that there was an argument to which he needed to reply, not just an 'opinion'. There is no such argument, and Hume's implicit account of the causal principle is more like Kant's own than Kant had any reason to suspect.[9]

Both Kant's dependence on Beattie's interpretation and the inaccuracy of Beattie's account have been disputed. Paul Guyer suggests that the content of Beattie's analysis of Hume would not have significantly added to Kant's interpretation.[10] Moreover, Manfred Kuehn has argued that Beattie presents an accurate account of Hume's causality.[11] These reservations concerning the inaccuracy and significance of Beck's analysis of Hume suggest that the factors that contributed to Kant's interpretation are found elsewhere.[12]

Scholarship discussing Hume's influence on Kant has generally focused on Kant's treatment of Hume's account of causality. In opposition to this view, Eric Watkins argues that Kant cannot be attempting to refute Hume's account of causality because the differences between their philosophies are so vast that they 'share no neutral philosophical vocabulary that would allow Kant to formulate a refutation of Hume on Hume's own terms'.[13] For Kant to be offering a refutation of Hume, there would need to be shared principles between their theories that were translatable from one theory to the other. However, the criterion that Watkins adopts for refutation are far too narrow[14] and would exclude most philosophical refutations except those that reveal an inconsistency in an opponent's position.[15] While Watkins's account of refutation is potentially too restrictive, he convincingly illustrates how Kant's treatment of Hume should not be regarded as an immanent refutation aimed at demonstrating Hume's falsehood on his own terms. Watkins develops this argument to show that Kant scholarship has over-exaggerated Hume's influence on Kant. Against Watkins, I argue that Hume exposed Kant to problems that inspired many central elements of transcendental idealism. Kant's separation of things in themselves from appearances, and the argument for synthetic *a priori* truths, were Kant's attempt to overcome the limitations of Hume's scepticism.

2.1.2 Kant's misunderstanding of Hume's relations of ideas and matters of fact

Kant's discussions of Hume in the first and second *Critiques* and the *Prolegomena* indicate that Kant regarded Hume's treatment of relations of ideas as the primary obstacle to the completion of Hume's own sceptical method. Moreover, Kant regarded his philosophy as correcting this oversight. I argue that Kant's misunderstanding of Hume's conception of relations of ideas meant that Kant did not recognise the similarity between this aspect of Hume's philosophy and his own development of the synthetic *a priori*.

According to Kant, Hume's scepticism was only directed at the relation between matters of facts and metaphysics.[16] The metaphysical principle towards which Hume directed his scepticism was causality; he asserted that '[a]ll reasonings concerning matters of fact seem to be founded on the relation of *Cause and Effect*'.[17] This scepticism drew attention to the insufficient justification for establishing deductive knowledge of future events from the occurrence of past events. Hume argued that it was impossible to produce any rational argument offering proof that future events will resemble past events. The statement 'the sun will not rise tomorrow' is no less intelligible or contradictory than its alternative. It is no less intelligible because it is impossible to offer any rational justification for the belief that one event follows another with causal necessity. According to Hume, 'when we reason *a priori*, and consider merely any object or cause ... it never could suggest to us the notion of any distinct object, such as its effect; much less, show us the inseparable and inviolable connection between them'.[18]

It is helpful to consider Kant's distinction between the justification (*quid juris*) for the concepts that we employ, and the fact (*quid facti*) that we do use these concepts.[19] Hume is not arguing that we do not use the concept of causality (*quid facti*); rather, he is arguing that we lack rational justification for our use of this concept (*quid juris*). Hume explains that we derive our belief in causality from custom or habit. Custom allows us to infer that one event will follow another without presuming to have offered rational justification for this propensity. According to Hume:

> Wherever the repetition of any particular act or operation produces a propensity to renew the same act or operation, without being impelled

by any reasoning or process of the understanding, we always say, that this propensity is the effect of *Custom*. By employing that word, we pretend not to have given the ultimate reason of such a propensity.[20]

Kant regarded Hume as having performed a service to philosophy by demonstrating that causal necessity could not be justified metaphysically.[21] However, Kant argued, Hume did not extend his scepticism far enough because of his distinction between relations of ideas and matters of fact. Kant understood Hume's relations of ideas as logical truths, which are known independently from experience. Hume apparently relegated mathematics to a pure analytic discipline devoid of any input from experience. Kant regarded himself as developing Hume's philosophy because he thought that Hume had protected mathematics from scepticism. According to Kant, it was as if Hume said that '[p]ure mathematics contains only analytic propositions, but metaphysics contains synthetic propositions *a priori*'.[22] Kant does not mean that Hume thought metaphysics contained synthetic *a priori* propositions in the sense of establishing transcendental necessary conditions for the possibility of experience. Rather, I suggest he thought that Hume's denial of the possibility to derive necessary causal relations from our inductive experience of objects revealed the impossibility of establishing *a priori* knowledge about objects independent of experience. For Hume, any *a priori* justification for metaphysical principles, such as causality, was inconceivable. Hume proclaimed such principles were nothing but sophistry and illusion and should be committed to the flames.[23] Kant wholeheartedly supported Hume's criticism of metaphysics based on the presuppositions of Hume's philosophy. According to Kant, '[w]hen Hume took the objects of experience as things in themselves (as is done almost everywhere), he was quite correct in declaring the concept of cause to be deceptive and a false illusion'.[24] This supports Watkins's claim that Kant could not be offering a refutation of Hume on his own terms, at least in relation to Hume's account of causality. Kant agreed with Hume's criticism on the condition that Hume's philosophy did not distinguish between appearances and things in themselves.

The essential limitation of Hume's philosophy was that he did not question whether our experience of objects is directly related to objects as they are in themselves. Hume appealed to the power of habit to explain the relationship between appearances and things in themselves. According to Hume, 'without any reasoning … we always suppose an external universe, which depends not on our perception, but would exist, though we and every sensible creature were absent or annihilated'.[25] This reinforces the opposition between reason and habit within Hume's philosophy. Any appeal to certainty regarding either the objects independent of experience or the necessary causal connections between these objects lacks rational justification and therefore is based on habit. Reason must remain silent on such issues as '[t]he mind has never anything present to it but the perceptions, and cannot possibly reach any experience of their connexion with objects. The supposition of such a connexion is, therefore, without any foundation in reasoning'.[26] The inability to either refute or justify this appeal to instinct is the embarrassment of philosophy. Philosophy cannot 'plead the infallible and irresistible instincts of nature: for that led us to quite a different system … And to justify this pretended philosophical system … exceeds the power of all human capacity'.[27]

Kant intended to show how philosophy could overcome this embarrassment by reorientating our enquiry towards explaining how the objects of experience are inseparable from the human faculties of sensibility, understanding and reason. The ambitious aim of transcendental idealism was to dispel any appeal to ignorance. Kant argued that experience must be understood in relation to the faculties that are necessary for the possibility of experience. This made it possible to reject Hume's appeal to rational ignorance regarding the human propensity towards the belief of causal necessity derived from habit. If we correctly understand the limits of knowledge in relation to these faculties, then we can explain why we are unable to answer certain questions because they transcend the limits of possible knowledge. Kant does not appeal to habit to explain the regularity between the appearance of objects and objects as they are in themselves; rather, he argues that the appearance of the object could not relate to the thing in itself. According to Kant:

> The strictest idealist cannot demand that one prove the object outside us … corresponds to our own perception. For if there were such a thing, then it still could not be represented and intuited outside us … The real outer is thus actual only in perception, and cannot be actual in any other way.[28]

The similarity between Kant and Hume is striking, yet Kant argues that the inability of reason to justify the relation between appearances and things in themselves does not show the deficiency of reason in relation to habit, but rather shows the illegitimacy of the extension of habit on these matters. For any speculative matter that transcends possible experience, 'the very same concept that puts us in a position to ask the question must also make us competent to answer it'.[29] In this sense, Kant's philosophy was aiming to extend the principles of Hume's philosophy and transform it into a systematic account of knowledge. In contrast, Hume approached philosophy merely as an academic exercise; he described how reason led him to 'philosophical melancholy and delirium' that was remedied by engaging in social activity. According to Hume, 'after three or four hour's amusement, I wou'd return to these speculations, they appear so cold, strain'd, and ridiculous, I cannot find it in my heart to enter into them any further'.[30] We can assume that Kant would have read this passage as it is from the chapter of the *Treatise* that was published in German in 1771. Kant's extension of Hume demonstrated that these were not merely academic concerns. According to Guyer, 'Kant does not refute Hume's academic worry about induction, but rather replaces it with a substantive theory of the imperfection of our knowledge of natural law'.[31]

Kant regarded Hume as coming closest to discovering the possibility of synthetic *a priori* truths.[32] In this context, he is referring to the use of synthetic *a priori* statements as necessary conditions for the possibility of experience, rather than exposing the illegitimate way that they were used to establish metaphysical arguments previously mentioned. For Kant, Hume did not realise the difference between the conditions of objects of appearance and objects in themselves because of his treatment of mathematics. He argued that Hume protected mathematics from his scepticism by regarding it as analytically and logically true or belonging to relations of ideas. According to Kant, if Hume had not done this:

> he would have to accept that mathematics is synthetic as well. But then he would by no means have been able to found his metaphysical propositions on mere experience, for otherwise he would have to subject the axioms of pure mathematics to experience as well, which he was much too reasonable to do.[33]

One of Kant's intentions for transcendental idealism was to preserve Hume's scepticism against traditional or dogmatic metaphysics while simultaneously establishing an alternative ground for certainty in relation to the necessary conditions for the possibility of experience. Kant regarded Hume's scepticism as potentially more far-reaching than Hume had apparently considered, stating: 'Hume's empiricism in principles also leads unavoidably to skepticism even with respect to mathematics and consequently in every *scientific* theoretical use of reason.'[34]

Kant thought that if Hume had realised the implications of his scepticism for mathematics, then this would potentially have led Hume to reorientate his account of causal necessity towards the discovery of a synthetic *a priori* ground in relation to appearances. Hume would have realised that mathematics also required a ground rather than merely assuming it was logically certain. The only way that Hume could have avoided mathematics becoming susceptible to scepticism, which he had deployed to great effect against the rational justification for causal necessity, would be to distinguish the conditions of knowledge of appearance from the conditions of knowledge of objects in themselves. Hume would have been able to simultaneously establish the certainty of mathematics and uphold his scepticism towards causal necessity pertaining to objects in themselves. However, if Hume had realised that mathematical knowledge could secure certainty in relation to experience, then he might have realised how knowledge of causal necessity could also be established in relation to appearances. These steps were precisely how Kant transformed the problem of induction, exposed by Hume, into the foundations for the system of transcendental idealism. Kant's comments on Hume suggest that he perceived a strong similarity between transcendental idealism and Humean scepticism. Transcendental idealism exposed the underlying assumptions that had evaded the scope of Hume's sceptical gaze. Hume had not distinguished appearances from things in themselves and he had apparently regarded relations of ideas as analytic and *a priori* truths.

Kant misunderstood Hume's account of relations of ideas. In both the *Enquiry* and the *Treatise* Hume did not regard mathematics as analytic in the sense that it was justifiable independent of its empirical demonstration. Kant assumed Hume's distinction between analytic and synthetic justifications was a logical

distinction, whereas Hume regarded it as a psychological distinction. To consider something as contradictory in the sense that it
is psychologically unimaginable does not entail that it is logically
impossible. Logical necessity is independent of demonstration,
whereas psychological necessity requires demonstration. Donald
Gotterbarn argues that both establish the same degree of certainty:
'[t]o claim that the certainty appropriate for these relations is
psychological rather than logical has no effect on the degree of cer-
titude appropriate to these relations'.[35]

Beck suggests that Kant misunderstood Hume because he had
not read the *Treatise*.[36] If he had, then 'he would have found
Hume tacitly admitting a class of intuitively and demonstratively
necessary relations of ideas which are not testable by the logical
law of contradiction'.[37] Against Beck's account, this distinction is
also evident in the *Enquiry* as Hume 'criticizes argumentation by
mere definition as "sophistry" and contrasts such arguments with
mathematical and demonstrative reasoning'.[38] Mathematics differs
from sophistic arguments because it depends on demonstration
and enquiry. Hume considers the following example of a sophistic
argument: '*that where there is no property, there can be no injustice*, it is only necessary to define the terms, and explain injustice
to be a violation of this property'.[39] In this case, the demonstration of the correctness of the statement merely depends on defining
terms in a circular way. In contrast, mathematics and geometry are
verified through experience. They are not matters of fact because
our knowledge of mathematical and geometric laws is not derived
from repetition and habit. In summary, if the distinction between
matters of fact and relations of ideas were a logical distinction,
then it would not be possible to explain Hume's further distinction
between relations of ideas and sophistic arguments.

Hume's account of relations of ideas as a psychological necessity
is surprisingly similar to Kant's account of synthetic *a priori* judgements. Both transcendental idealism and Humean empiricism were
concerned with the possibility of establishing *a priori* certainty,
which is derived synthetically.[40] Relations of ideas demonstrate
their *a priori* necessity because it is psychologically inconceivable
that they could be otherwise. It is a process of reasoning by which
the synthetic demonstration of a proof, axiom or theory entails its
certainty. These proofs are not justified by the repetition of events
experienced as regularities derived from habit or custom. The

certainty of this kind of knowledge is derived from what Hume terms the 'operations of thought'. This is strikingly similar to Kant's conception of synthetic *a priori* judgements.[41]

In summary, Kant's misunderstanding of Hume's distinction between relations of ideas and matters of fact was instrumental for the development of Kant's critical philosophy. Hume did not protect mathematics from his scepticism in the sense that he merely assumed that mathematics was immune to the problems identified for matters of facts. Like Kant, Hume thought that mathematics was justified in accordance with synthetic experience in cases where it was possible to demonstrate *a priori* certainty. Despite the underlying similarity between Hume's conception of mathematics and Kant's account of the synthetic *a priori*, this did not cause Hume to reassess his entire philosophy as Kant had suggested it would.

2.2 Interpretations of Hume's philosophy

According to the sceptical realist interpretation of Hume, he was only sceptical of laws at the epistemological level, but he nevertheless believed in the ontological existence of laws. I examine how Hume's appeal to pre-established harmony offers support to this interpretation. Pre-established harmony, originally formulated by Leibniz, suggested there was no direct causal relation between physical and mental entities. The apparent cause and effect relations we find in nature have actually been pre-determined by God. Hume's appeal to pre-established harmony stands in tension with his empirical scepticism. Comparing Kant's interpretation of Hume with the sceptical realist interpretation consolidates the importance of Hume's philosophy for Kant and exposes important ways that Kant diverges from Hume. Sceptical realism capitalises on the ambiguity of Hume's appeal to laws independent of experience, whereas transcendental idealism demonstrates why knowledge of such laws is impossible.

2.2.1 The sceptical realist interpretation of Hume

Kant's interpretation of Hume must be distinguished from the sceptical realist interpretation of Hume, which suggests that Hume was an ontological realist about the existence of causal powers and external objects. This realism is sceptical because, despite positing the existence of causal powers and external objects, it also claims that we cannot know anything about them because of our epistemic limitations. Galen Strawson argues that Hume was merely an epistemological – not an ontological – anti-realist about causal laws. According to Strawson, the ontological anti-realist interpretation of Hume conflates Hume's epistemological claim that '[a]ll we can ever know of causation is regular succession' with the ontological claim '[a]ll that causation actually is, in the objects, is regular succession'.[42] Strawson appeals to various references that Hume makes to the secret springs and principles of nature in support of Hume's belief that such entities existed. Hume merely intended to show that we could not justify these principles from an epistemological standpoint. According to Kenneth Richman, '[a] sceptical realist about some entity is realist about the entity's existence, but agnostic about the nature or character of that thing because it is epistemically inaccessible to us in some non-trivial way'.[43]

Hume describes the correspondence between custom and nature as 'a kind of pre-established harmony between the course of nature and the succession of our ideas'.[44] This supports the sceptical realist interpretation insofar as it demonstrates Hume's belief in the existence of the external world. Hume's appeal to pre-established harmony is an implicit appeal to Leibniz's philosophy. Pre-established harmony is Leibniz's solution to the problem of the causal interaction between the mind (or soul) and the body. For Leibniz, there is no causal interaction between these substances. Their apparent reciprocal causal influence is a pre-established harmony maintained by God. The soul and body are in conformity with each other, as '[t]he soul follows its own laws and the body also follows its own; and they agree in virtue of the harmony pre-established between all substances, since they are all representations of a single universe'.[45] Hume's appeal to pre-established harmony is in tension with his scepticism as it is beyond empirical justification. According to Guyer, this appeal is sincere and demonstrates Hume's deviation from 'true scepticism' in the *Enquiry*. In contrast, Juliet Floyd argues this passage is 'drenched with characteristic Humean irony'.[46]

If Hume's statement is saturated in irony, then the question of the correspondence between custom and nature remains unanswered. The sincerity of Hume's commitment to pre-established harmony is less important than the underlying problem. That is, Hume appealed to pre-established harmony because he could not conceive of a better explanation for the correspondence of our ideas with the powers of nature. According to Hume, nature has 'implanted in us an instinct, which carries forward the thought in a correspondent course to that which she has established among external objects; though we are ignorant of those powers and forces, on which this regular course and succession of objects totally depends'.[47] The habitual or instinctual belief in causality is so powerful that we cannot seriously doubt or suspend it. For Hume, there is no rational basis for explaining the conformity of objects in nature with custom or habit. Hume's blind faith in habit presents his scepticism in a very different manner. Hume uproots our rational justification for pre-established harmony, yet he maintains that we can continue believe in pre-established harmony, and the existence of objects independent of experience, and their causal relations, on the basis of habit or custom. According to Guyer:

> He [i.e., Hume] certainly does not argue that we should suspend our belief in causality, as a genuine sceptic might; on the contrary ... he claims that 'no reasoning or process of the thought and understanding' is capable of preventing the operation of our natural instincts to form causal belief.[48]

Guyer argues that Kant's awakening from his dogmatic slumber might not have been caused by Hume's scepticism but rather his dogmatism, that is, his 'assumption of the good fit between our faculties and the world around us'.[49] In opposition to Guyer, the sceptical realist interprets Hume's belief in the good fit of our faculties with real powers in the world as Hume's avoidance of dogmatism. The sceptical realist opposes any interpretation of Hume that 'dogmatically denies the existence of real powers, or dogmatically affirms that causation is nothing but constant conjunction'.[50] The sceptical realist avoids the charge of dogmatism associated with denying the existence of ontological powers, but then they dogmatically posit the existence of ontological powers even though they are beyond the remit of knowledge.

For Strawson, Hume's appeal to pre-established harmony makes it possible for Hume to establish a connection between causality at the epistemological and ontological levels. According to

Strawson, 'the Imagination works the merely regular-succession notion of causality in the objects into a truer notion of causation in the objects, a notion of Causation'.[51] Strawson explains that Hume could not present his account in these terms because of his empiricism towards conceptual content. He argues that Hume's account entails a more modest version of this argument. According to Strawson, Hume 'can, certainly, say that it's part of the wisdom of nature to make us rely implicitly on inductive procedures, and reason about matters as if we knew that the world was governed by necessary causal laws'.[52] For Strawson, Hume's appeal to the wisdom of nature amounts to our innate capacity to deploy concepts, which could be the product of God or evolution.

There are two problems with Strawson's account. First, even if Hume did regard the wisdom of nature as something like the product of evolution[53] or God, this would not explain our justification to comprehend nature as if it were governed by necessary causal laws (*quid juris*). It would merely show that we do comprehend nature as such (*quid facti*). Second, Strawson's explanation of Hume's modest claim is strikingly similar to Kant's account of nature as ordered by regulative principles projected onto nature by the subject. Against Strawson, Kant's philosophy reveals why Hume lacks the conceptual apparatus to justify the conclusion that we must regard nature as if universal and necessary laws governed it. The account of reason required to justify this conclusion is insufficient in Hume's philosophy. Kant's account of reason provides the framework to achieve what is impossible for Hume. According to Allison:

> The assumption of nature as exhibiting a systematic unity enables Kant to explain how reason makes it possible to account for what, according to Hume, was possible only on the basis of custom, namely, the inference from the observed to the unobserved ... Reason accomplishes this by 'projecting an order of nature', which licences the inferences from the law-like regularities discovered in the limited portion of experience with which we happen to be acquainted with at any point in time to experience as a whole.[54]

Kant established the rational framework to understand nature *as if* it is governed by universal laws. Kant mobilised reason into a faculty that is essential for our ability to view appearances *as if* they are law-governed and systematic, whereas Hume denied that reason could play any role in this process. Kant's interpretation diverges from

Strawson because Kant is concerned with exposing the limitations of Hume's philosophy. Strawson claims that Hume believed causal laws are necessary connections between objects that exist independent of experience, whereas Kant is concerned with understanding causality as a transcendental condition for the possibility of experience. Kant revered Hume's philosophy for revealing the limits of metaphysics and its dogmatism in the transgressions of these limits, but the principle of doubt is an acidic principle that is equally corrosive to both transcendent and transcendental claims. To this effect, Kant described Hume's philosophy as 'the **censorship** of reason'.[55] Hence, Kant describes scepticism as a mere resting-place for reason, but 'not a dwelling-place for permanent residence; for the latter can only be found in complete certainty, whether it be one of the cognition of objects themselves or of the boundaries within which all our cognition of objects is enclosed'.[56] Hume's influence on Kant was far more extensive than his scepticism towards the rational justification of causal necessity. According to Watkins, 'Kant does not so much present detailed *arguments* against skepticism as provide an *orientation* that views skepticism as an unstable position'.[57]

Kant explains the development of philosophy as analogous to the development from childhood to adulthood. Dogmatic metaphysics corresponds to the childhood of philosophy. Scepticism marks the intermediate stage. Unsurprisingly, the complete maturation of philosophy is marked by the *critique*, rather than the *censorship*, of reason. This description of the lineage of philosophy further emphasises Kant's distinction between transcendental realism and transcendental idealism. Hume criticised transcendental realist metaphysics because it depended on knowledge that went beyond experience, but his solution was merely to assume the presuppositions of transcendental realism were correct independent of any rational justification for them. Both the sceptical realist and Kantian interpretations of Hume agree on this point. In contrast to Strawson, Kant's intention was not to provide an interpretation of Hume; he surely would have opposed Strawson's attempt to validate Hume's belief in transcendent metaphysical entities. Instead, Hume was a 'springboard' for the development of Kant's philosophy as he revealed the inability of reason to infer any metaphysical claims from experience pertaining to objects in themselves.

The analogy of the history of metaphysics as a developmental process is another example of how Kant appealed to physiological principles to elucidate his theory. In the previous chapter, I examined Mensch's argument that Kant appealed to 'nature as a model by which to interpret reason'.[58] This was in relation to Kant's comments on the 'epigenesis' of reason.[59] Kant borrowed his account of epigenesis from the physical sciences to explain the emergence of the architectonic system of transcendental idealism. He transformed epigenesis from an explanation of the origins of entities in nature to an explanation of the origin of the system of reason.[60] This enabled Kant to argue that the emergence of reason did not require further explanation; or, at least, that it was not possible for finite rational beings to establish such an explanation.[61]

In summary, there is sufficient evidence from Kant's treatment of Hume to consider Kant inspired criticisms against the sceptical realist interpretation. Kant regards transcendental idealism as placing reason at the centre of human activity as a way of extending the scope of scepticism. In contrast, Strawson's sceptical realist interpretation emphasises Hume's belief in entities that cannot be rationally justified in accordance with Hume's empiricism. Moreover, Strawson's interpretation of Hume as approaching the world *as if* it were governed by necessary causal laws has a greater resemblance to the Kantian, rather than Humean, account. This greater resemblance to the Kantian account is indicative of how Kant developed his philosophy in response to the problems raised by Hume.

2.2.2 Transcendental idealism as developing from Humean empiricism

One significant way that Kant's philosophy differs from Hume's is that Kant argues that space and time cannot be regarded as part of the world independent of experience, instead they are only ever known in the context of experience. Kant refers to space and time as *Anschauung*, which is conventionally translated as 'intuition'.[62] Kant further divides each form of intuition (space and time) into pure and empirical kinds. Kant's arguments for intuitions focus on two features: they are synthetic *a priori* and non-discursive. Kant aims to show that intuitions are distinct from concepts in order to show that experience requires both intuitions and concepts, and neither can be reduced to the other. The uniqueness of intuitions depends on the demonstration of their non-discursivity or non-conceptuality. Kant argues that time is *a priori* on the following grounds:

Time is not an empirical concept that is somehow drawn from an experience. For simultaneity or succession would not themselves come into perception if the representation of time did not ground them *a priori*. Only under its presupposition can one represent that several things exist at one and the same time (simultaneously) or in different times (successively).[63]

All experience is either successive or simultaneous, but if knowledge of this were derived from experience, we could not know these are necessarily the only ways that experience can be presented to us. We could only make the inductive statement that every experience up to this point has been either successive or simultaneous. If Kant's argument is correct, then it only demonstrates that the successive or simultaneous modes of time are *prior* to experience because they are conditions that make experience possible. However, it does not offer sufficient proof for the *a priority* of time.

Kant's second argument attempts to offer additional justification for the *a priority* of time. It takes the form of a thought-experiment aimed at demonstrating that time cannot be derived from experience. Kant asks us to consider whether it is possible to think of an absence of time. He asserts, 'one can very well take the appearances away from time ... but time itself (as the universal condition of their possibility) cannot be removed'.[64] As the abstraction of the object does not entail the abstraction of time, Kant concluded that time must be the condition of the possibility of the experience – rather than a property – of objects in themselves. Both time and space are universal *a priori* conditions for the possibility of experience, without them experience would not be possible.

Allison contrasts Kant's and Hume's accounts of time by asking if events can be directly perceived, or does experience 'require some kind of interpretive act guided by an a priori rule?'[65] For Hume, no such rule is necessary, but this leads to further problems. Allison explains that Hume cannot adequately account for the perception of successive events or successive states of an object. He considers Hume's example of hearing five notes played on a flute.[66] Hume argues that time does not form a sixth impression; however, Hume's account requires an explanation for how the mind can contain the idea of the five impressions. According to Allison:

he [i.e., Hume] has no way to explain how, on the basis of five successive note perceptions, we can arrive at a perception *of the succession* of five notes. What we now learn, with the help of Kant, is that such a perception ... requires what I have termed an act of interpretation governed by an a priori rule ... [T]he Humean mind enters the story one step too late.[67]

The Humean mind enters the story one step too late because it assumes the experience of the succession of time requires no special explanation. Moreover, Hume's account means that he can only explain the ordering of appearance as an additional impression. Kant would agree with Hume insofar as the order or manner of appearances is not sufficient to establish that time is an additional impression. That is, time is not something additional to any particular experience. However, for Kant, this serves as justification for conceiving of time as an *a priori* condition for the possibility of experience in general. From the perspective of transcendental idealism, Hume cannot explain how we first become aware of the succession of the five notes. Allison asserts that 'unless time were presupposed as the medium or framework in which this succession is perceived, one could not be conscious of the notes as successively occurring in it'.[68] The shift in Hume's empiricism to Kant's transcendental idealism is not as counter-intuitive or far-fetched as it is often argued to be. The Humean empiricist is unable to identify any conditions for the possibility of experience, which is a fundamental task of transcendental idealism.

Guyer has raised a criticism against transcendental idealism on the basis that it could be possible for space and time to be properties of objects in themselves, and we become empirically aware of space and time in the early stages of development before any formation of memory is possible.[69] In this case, space and time would only seem to be *a priori*, but they would in fact be *a posteriori* as they are derived from contingent experience. If space and time are contingent, then it is impossible in principle to rule out the possibility that appearances could be represented in other ways. It is also impossible to declare that objects independent of experience are necessarily devoid of space and time. Hence, Guyer argues that Kant does not have sufficient grounds to demonstrate the necessity or the *a priority* of space and time as the only forms of intuition. Moreover, he argues that Kant concedes this when he asserts that no further ground can be offered 'for why space and time are the sole forms of our possible intuition'.[70]

Guyer's interpretation of transcendental idealism leaves it defenceless against the criticism of the *a priori* necessity of space and time. In his view, Kant's arguments for the *a priority* of space and time is established by denying that properties extend to objects in themselves, but Kant requires 'some *independent reason* to hold that those epistemic conditions *could not also be* properties of objects'.[71] Allison argues that Kant's position is the reverse of this claim; we lack justification to attribute space and time to objects in themselves because space and time are only ever epistemic conditions. Allison's interpretation takes a more sympathetic stance towards transcendental idealism:

> [T]he issue currently before us is not whether epistemic conditions might somehow be *satisfied* by things as they are in themselves, but rather whether the representations that function as such conditions (in this case the forms of sensibility) might be *derived from* an experience of things so considered.[72]

Guyer adopts an essentially Humean stance toward space and time, namely that we cannot know whether they are properties of things in themselves or experience. This sets Kant two incredibly difficult tasks. First, he must provide sufficient evidence for why space and time are not properties of things in themselves. Second, he must explain space and time as intuitions belonging to the human faculty of sensibility. Allison correctly shows that the first task is not necessary or even relevant for the integrity of Kant's project. The previous analysis of Hume explored how starting from the assumption that space and time are properties of objects creates its own problems about how we organised sense-contents into experiences such as notes played on the flute.

The differences between their accounts stems from a more fundamental disagreement between the two-world and two-aspect interpretations. Guyer argues that the overwhelming prevalence of the two-world account in philosophy at Kant's time is strong evidence in support of the two-world interpretation of Kant; 'of course he [i.e., Kant] held a "two-object" view: everyone (except Berkeley) did, though few would have agreed with Kant's reassignment of spatio-temporal properties from ordinary objects to representations'.[73] The idea that space and time were re-assigned from objects to representations implies that Kant needed to demonstrate that these properties could not belong to things in themselves. On the other hand, if transcendental idealism was revealing the insufficient justification for claiming that these were properties of objects, then all that is required is to demonstrate how space and time are conditions of the representations of objects.

The two-aspect interpretation concedes that it is not possible to provide independent reasons to demonstrate that space and time are not conditions of objects in themselves, but it argues that this question is beyond the scope of knowledge. In other words, the presupposition that space and time should relate to things in themselves, rather than appearances, is an assumption lacking sufficient justification. In this context, the two-aspect interpretation emphasises Kant's argument that space and time can only be known as necessary conditions for the possibility of experience if they are understood as emerging from the subject.

The complexity of the relationship between appearances and things in-themselves highlights the difficult relationship that Kant's philosophy has with biology. There has been a renewed interest in the notion of biological autonomy, which specifically appeals to aspects of Kant's philosophy as offering support to their account.[74] According to the two-world interpretation, we can assert that our autonomy is not derived from nature as nature is spatio-temporal and governed by efficient causation, whereas autonomy requires that we view ourselves as free noumenal agents. Kant explains that an individual is subject to laws when considered as an appearance that do not apply when that individual is treated as a thing in itself.[75] According to Allen Wood, we are 'simultaneously free and causally determined because we belong to two worlds'.[76] For the sake of this argument, let us assume this interpretation is correct. It exposes a potential tension for philosophers of biology who appeal to Kant's philosophy. They will not be satisfied with any account of freedom and autonomy that is grounded in a world other than nature; biological autonomy is essentially a property of biological entities.

The two-aspect interpretation is also problematic for biological conceptions of freedom. The two-aspect view can only allow us to conceive of our freedom as abstracted 'from the causal determinations of our actions in space and time'.[77] This is not the same as claiming that the source of causal determination of our actions could be other than spatial and temporal. For Allison, practical reason functions by developing and projecting maxims that 'guide conduct by framing an order of ends or ought-to-bes'.[78] This projective capacity of practical reason is just like the projective capacity of theoretical reason, both are regulative principles that guide enquiry; '[l]ike its theoretical analogy, this activity is an expression of spontaneity of reason because it goes beyond what is dictated by sensible data'.[79] This leads Allison to distinguish rational necessity and causal necessity as only the latter is derived from sensibility. In short, reasons are not causes. According to Allison's interpretation, freedom is not proven, but presupposed. Freedom is acting *as if* we are not like objects of experience that are causally determined in space and time. This is potentially unsatisfactory for contemporary accounts of freedom in philosophy of biology as it relegates freedom to a mere presupposition.

Kant relates his account of freedom and causality back to his interpretation of Hume. According to Kant:

> The concept of an empirically unconditioned causality is indeed theoretically empty, (without any intuition appropriate to it) but it is nevertheless possible and refers to an undetermined object; in place of that, however, the concept is given significance in the moral law and consequently in its practical reference.[80]

Kant agrees with Hume that the idea of an unconditioned causal necessity is not something that could be found within experience, and therefore it is theoretically empty. The idea of unconditioned causal necessity lacks any corresponding intuition, which is a necessary component for the possibility of experience. Thus, it is impossible to have any experience of unconditioned causal necessity. Hume argued that causation cannot be known beyond habit or custom. In contrast, Kant suggests another way of establishing unconditioned causal necessity from the perspective of practical reason. Crucially, practical reason does not directly correspond to intuitions, and therefore the 'objects' of practical reason are fundamentally different from objects of theoretical reason. According to Kant, '[b]y a concept of an object of practical reason I understand the representation of an object as an effect possible through freedom'.[81] Objects of practical reason have very little in common with objects of theoretical reason. They cannot be objects of experience as they are made possible through the presupposition of freedom, which is not given in experience.

In a sense, Kant reverses Hume's rejection of knowledge of unconditioned causal necessity. It does not follow that there are no possible alternative ways to establish causal necessity from Hume's argument that causal necessity cannot be known within experience. Transcendental idealism opened the space for the investigation into the necessary conditions for the possibility of morality, but this is based on our ability to create a universal maxim in accordance with the categorical imperative. From this perspective, Hume's scepticism towards causal necessity plays an essential role for both Kant's theoretical and practical philosophy.

In summary, Kant's separation of intuitions from things in themselves made it possible for him to establish space and time as necessary transcendental conditions for the possibility of experience alone. Kant agreed with Hume that it is not possible to know causal necessity between objects in themselves from the perspective of theoretical reason. However, this highlights a distinction between theoretical and practical reason within Kant's philosophy, as he argues we can, and should, deploy practical reason to generate universal moral maxims. Our ability to generate these maxims is compatible within Kant's broader account of theoretical reason because his denial that knowledge of causal necessity relates to objects in themselves allows for the possibility of freedom. The relationship between theoretical and practical reason is far from straightforward, which is clear from our discussion of alternative interpretations of Kant. Regardless of the interpretation that guides our understanding of Kant's philosophy, significant differences emerge between the presuppositions of Kant's philosophy and the presuppositions of science.

2.3 Metaphysical arguments regarding the existence of laws

In this section, I apply Kant's philosophy to ontological accounts of the laws of nature in contemporary philosophy of science. The two philosophies under consideration are Cartwright's account of laws as 'nomological machines' and Bhaskar's transcendental realist account of laws. For Bhaskar, the existence of universal laws is a precondition for the possibility of scientific practice, whereas

Cartwright argues there is little empirical evidence to support the existence of laws beyond their manifestations derived from scientific experiments. After outlining both accounts, I argue that combined they can be presented in the form of a Kantian mathematical antinomy. Both presuppose that the conditions of experience can be applied to metaphysical arguments that pertain to entities beyond appearance. Despite the obvious divergence between their accounts regarding the existence of universal laws, there is also a significant point of similarity. I argue that both accounts presuppose that it is possible to regard science as unified, albeit in different ways. Bhaskar locates this unity in universal transfactual laws of nature, whereas Cartwright identifies this unity at the level of causal capacities, rather than laws. Both accounts claim to be offering transcendental arguments to justify their positions, yet their arguments reach incompatible conclusions, which Kant argues should not be possible. I argue that this is because they overstep the boundaries of knowledge identified by transcendental idealism, which maintains that appearances are both empirically real and transcendentally ideal.

2.3.1 Bhaskar's transcendental realism and Cartwright's nomological pluralism

Cartwright and Bhaskar emphatically oppose Kant's critical philosophy. Cartwright criticises transcendental idealism on the basis that it imposes an unrecognisable image of reality. According to Cartwright, transcendental arguments:

> appear in the clean and orderly world of pure reason as refugees with neither proper papers nor proper introductions, of suspect worth and suspicious origin. The facts which I take to ground objectivity are similarly alien in the clear, well-lighted streets of reason, where properties have exact boundaries, rules are unambiguous, and behaviour is precisely ordained.[82]

Cartwright's reflection on the 'alien' nature of transcendental arguments highlights the differences between the transcendental methodology and the methodologies adopted by many contemporary philosophers. The notion that Kant's streets of reason are 'well-lighted' with 'exact and unambiguous boundaries' conflates the goal of transcendental idealism with its procedure. Kant's intention is to provide the basis for the possibility of any future metaphysics,[83] not to provide an exhaustive account of the content of this metaphysics. Cartwright's strong commitment to empiricism stands against the possibility of the foundations of transcendental idealism such as apodicticity and universality, as she argues there is little empirical justification for the assumption that nature conforms to these principles. In contrast, Bhaskar opposes transcendental idealism because he regards it as a continuation of empiricism.[84]

Bhaskar argues that Hume's empiricism entails the ontological denial of the existence of necessary laws of nature; 'the empiricist fills the [ontological] vacuum he creates with his concepts of experience. In this way an implicit ontology, crystallized in the concept of the empirical world, is generated'.[85] Any justification for scientific enquiry is denied by Hume's ontology because it denies any knowledge of the laws that are presupposed for such an enquiry.[86] Bhaskar argues that transcendental idealism is inadequate because it adopts the empiricist account of being as relative to experience, 'although transcendental idealism rejects the empiricist account of science, it tacitly takes over the empiricist account of being'.[87] Both transcendental idealism and empiricism are committed to empirical realism. Empirical realism is not inherently problematic; however, it becomes problematic if experience is regarded as an exhaustive account of the domain of human activity. The notion that experience provides an exhaustive account of reality does not accurately represent Kant's philosophy, as many aspects of human activity are not justified in accordance with empirical realism. For instance, Kant argues that the power of judgement and practical reason cannot be appropriately justified in relation to theoretical philosophy, but he is nonetheless committed to experience as the ground for theoretical knowledge and the theoretical deployment of reason. According to Kant, '[a]ll of our cognition is in the end related to possible intuitions: for through these alone an object is given'.[88]

Bhaskar's criticism correctly identifies the empirical realist implications of Kant's conception of theoretical reason. This has been overlooked by other criticisms of Kant's philosophy such as the equivocation between transcendental idealism and Berkeleyian idealism.[89] Kant distinguished his philosophy from Berkeley's on the basis that his philosophy provided the means for distinguishing truth from illusion. Only transcendental idealism could establish transcendental principles that were necessary for the possibility of experience. According to Kant:

> For *Berkeley* experience could have no criteria for truth, because its appearances (according to him) had nothing underlying them *a priori*; from which it then followed that experience is nothing but sheer illusion, whereas for us space and time (in conjunction with the pure concepts of the understanding) prescribe *a priori* their law to all possible experience, which law at the same time provides the sure criterion for distinguishing truth from illusion in experience.[90]

Kant avoided the implications of George Berkeley's empirical idealism by aligning his philosophy to Hume's empirical realism. Kant's response to Berkeleyian idealism is to show that knowledge can only arise where concepts apply to intuitions. Theoretical claims that lack any reference to intuitions cannot be justified because there is no way to demonstrate how these claims correspond to an object. In contrast to Kant, Bhaskar argues that being is necessary at both the epistemological and ontological levels. According to Bhaskar, 'we can allow that experience is in the last instance epistemically decisive, without supposing that its objects are ontologically ultimate, in the sense that their existence depends on nothing else'.[91] Bhaskar argues that ontology is distinct from epistemology insofar as the role of philosophy is to furnish science with an ontology that promotes the flourishing of science.[92] Ontology is a hand maiden to science for Bhaskar: '[o]ntology, it should be stressed, does not have as its subject matter a world apart from that investigated by science'.[93] In this way, Bhaskar demarcates the appropriate domain of ontological enquiry. Any philosophy that either pursues ontological enquiries independent from the sciences or rejects the possibility of a realist ontology of science is invalid. This is opposed to the approach towards the philosophy of science developed in this book, which examines the various ways that philosophers of biology have appealed to aspects of Kant's philosophy.

In the previous chapter, we discussed Kant's critique of transcendental realism. For Kant, transcendental realism entails that the conditions of knowledge of appearances are treated as the conditions of knowledge of things in themselves.[94] Bhaskar avoids this by separating the transitive and intransitive aspects of scientific enquiry. The transitive objects of science relate to the aspects of a theory that constitutes its empirical verification. These are the raw materials of science, which include the techniques and apparatus available to scientists. This refers to the process of constructing an environment that imposes an artificial closure on nature for the purposes of scientific enquiry. A scientific experiment requires that the conditions of experimental production and experimental closure are achieved. Experimental production entails that an experiment must trigger a relevant natural mechanism, experimental closure entails that all other variables are isolated. To put it in more Kantian terms, without experimental production, an experiment would be empty as no mechanism would be triggered. Without experimental closure, the experimenter would be blind to other mechanisms that could invalidate any results.[95] Bhaskar explains that the regularities established in an experiment do not generally reflect nature beyond these experimental conditions. According to Bhaskar:

> The aim of an experiment is to get a single mechanism going in isolation and record its effects. Outside a closed system these will normally be affected by the operations of other mechanisms ... so that no unique relationship between the variables or precise description of the mode of operation of the mechanism will be possible.[96]

The mechanisms discovered in a closed system or scientific experiment, which are transitive objects, do not entail that the regularity of these mechanisms extends to open systems. Recall that, for Hume, even if an effect were to follow a cause without exception, it would still not justify us to derive the necessary connection between these events from our experience of their constant conjunction. Bhaskar and Cartwright both agree that nature does not generally conform to this regularity beyond the context of scientific experimentation. They diverge because Bhaskar argues for the realism of laws as a rational condition for the possibility of science, whereas Cartwright argues that there is no empirical basis for the existence of universal laws.

For Bhaskar, knowing how these laws act in open systems independently from scientific enquiry is not a requirement for concluding that scientific enquiry is examining those same laws in closed systems. According to Bhaskar:

> The transcendental realist sees the various sciences as attempting to understand things and structures in themselves, at their own level of

being, without making reference to the diverse conditions under which they exist and act, and as making causal claims which are specific to the events and individuals concerned. And he sees this not just as a tactic or manoeuvre or mechanism of knowledge; but as according with the way things really are, the way things must be if our knowledge of them is to be possible.[97]

This is transcendental because it is an explanation of how scientific laws must be for the possibility of scientific knowledge; however, this certainty does not require empirical verification. We cannot know the actions of these laws independent of scientific enquiry in open systems more generally. Nevertheless, we must regard these laws as constantly acting irrespective of the absence of their regularity in nature. Bhaskar terms these laws transfactual statements. Transfactual statements refer to the lawful activity of nature independent from experience, and science is the contingent enquiry into these laws that is specific to human activity. In Bhaskar's terms, the former relates to intransitive objects of science, whereas the latter relates its transitive objects.

The intransitive objects of science persist independent of transitive scientific enquiry; '[w]e can easily imagine a world similar to ours, containing the same intransitive objects of scientific knowledge, but without any science to produce knowledge of them'.[98] It is important to distinguish the claim that intransitive objects are necessary for the possibility of science from his claim that the same intransitive objects are conceivable in a different world. This claim is problematic because it does not follow from the mere conceivability of another world, with the same intransitive objects of science, that these intransitive objects are necessary for all possible worlds. While the discovery of any scientific theory is contingent, the intransitive laws underpinning that discovery are necessary independent of experience.

The problem with this proof of intransitive aspects of science is that even if the absence of the intransitive aspects of science were unimaginable, this only demonstrates, at best, the psychological necessity – rather than the ontological necessity – of scientific laws independent of experience. Ontology posits the existence of metaphysical entities, whereas psychology posits such entities as necessary from the human perspective. The intransitive aspects of science allow Bhaskar to resolve the tension between necessity and contingency in the development of scientific practice. Bhaskar's claim that this amounts to a transcendental justification of laws independent of experience is insufficient. For Bhaskar, the practice of science requires the ontological necessity of intransitive laws, and it is the purpose of philosophy to furnish science with this ontology.

The essential difference between Bhaskar and Cartwright is that the former provides a rationalist argument for the existence of laws whereas the latter endorses a critical empiricism with respect to laws. Empirical evidence has generally demonstrated that laws are context-specific, rather than being fundamental and universal regularities that persist throughout nature. According to Cartwright, '[m]etaphysical nomological pluralism is the doctrine that nature is governed in different domains by different systems of laws not necessarily related to each other in any systematic or uniform way; by a patchwork of laws'.[99] A law is the nomological machine that is produced in the context of experimental conditions; '[t]o grant that a law is true … is far from admitting that it is universal – that it holds everywhere and governs in all domains'.[100]

Cartwright's account of nomological machines is similar to Bhaskar's account of the transitive objects of science as both describe the context-specificity of scientific research. Cartwright diverges from Bhaskar because she claims that the successes of scientific research demonstrates its context-specificity. She argues that some of the best examples of scientific progress focus on the context-specific background of capacities in nature. Aviation is a paradigm example of harnessing these capacities to produce regularity with great effect; this is one of many examples demonstrating our ability to construct entities 'to fit the models we know work. Indeed, that is how we manage to get so much into the domain of the laws we know'.[101] Understanding laws as context-specific does not require that these laws possess a predictive capacity beyond the context from which they are derived. On the contrary, nomological pluralism suggests that scientific research should be expanding our understanding of laws by directing research toward similar local models where we could reasonably predict that similar natural capacities might be present.

Although Cartwright considers her account to stand in opposition to the traditional conception of unified science, she does not exclude the possibility of a unified science at a more general level. Cartwright is only denying that we can understand this unity under a model that considers laws as fundamental and universal regularities. The empirical evidence to support this account of laws is generally very weak, not merely because of the problem concerning deductive knowledge of laws based on inductive inferences, but because science has demonstrated that nature rarely expresses law-like regularity. Newtonian mechanics was exceptional because it discovered law-like regularity and became a paradigm that both science and philosophy strove to achieve. However, such law-like regularity has rarely been discovered without imposing a plethora of *ceteris paribus* conditions. According to Dupré, 'far from knowing that these laws are universally true, we know that they are generally false'.[102] Cartwright is not suggesting that laws are false, merely that we falsely infer that laws continue to act independently from their scientific context. We must have reasons to suggest that any law extends beyond the nomological machine that demonstrated it. For Cartwright, this suggests the need to focus on the natural capacities that form the context of the law. According to Cartwright:

> I do not deny the unifying power of the principles of physics. But I do deny that these principles can generally be reconstructed as regularity laws. If one wants to see their unifying power, they are far better rendered as claims about capacities, capacities that can be assembled and reassembled in different nomological machines, unending in their variety, to give rise to different laws.[103]

Cartwright's appeal to capacities as the basis for the unity of science is diametrically opposed to Bhaskar's account. She reverses the relationship between experiment and laws on the one hand, and laws and unity on the other. Laws are the product of scientific experiments, and a more appropriate ground for a unified science focuses on the various ways that natural capacities can be assembled and reassembled.

2.3.2 A mathematical antinomy: Bhaskar and Cartwright

One way to explain the difference between the positions of Bhaskar and Cartwright is in terms of the metaphysical difference between rationalism and empiricism. Cartwright takes the position of empiricism and Bhaskar the position of rationalism. This allows us to formulate their arguments as a mathematical antinomy consistent with the form that Kant presents in the first *Critique*. The importance of the antinomies cannot be underemphasised. In a letter to Garve in 1798, Kant stated that the antinomies 'first aroused me from my dogmatic slumber and drove me to the critique of reason itself, in order to resolve the scandal of ostensible contradiction with reason itself'.[104] Notice that this is not the first time Kant had spoken about awaking from his dogmatic slumber. He used the same phrasing in the *Prolegomena* where he identified Hume, rather than the antimonies, as the source of his awakening. This is not an inconsistency in Kant's philosophy, according to James O'Shea central to both Hume's philosophy and the antimonies is the 'attempt to demonstrate the impossibility of achieving any sound rational arguments, independently of experience, for any synthetic a priori conclusions about reality'.[105]

Kant argues that the contradictions of reason discussed in the antinomies arise naturally and are indicative of the tendency of reason to overstep the boundaries of knowledge. Kant distinguishes between two types of antinomy in the first *Critique*. These are mathematical antinomies and dynamical antinomies. The former relates to cases where both sides of the contradiction depend on a shared assumption. Kant describes this as 'a synthesis of homogeneous things', whereas the latter is 'a synthesis of things not homogeneous'.[106] Kant continues to explain how mathematical antinomies require that both sides of the argument refer to a sensible condition, whereas dynamical antinomies relates to cases where one side of the argument refers to a sensible condition and the other refers to an intelligible condition. This means that there is no shared assumption for dynamical antinomies.

Kant's antinomies in the first *Critique* take the form of *reductio ad absurdum* arguments. This argumentative form offers justification for an initial position by assuming the opposite stance and revealing the internal inconsistency of the position, resulting in contradictory implications. As the opposite stance is contradictory, this offers support to the initial position as it is assumed as the only possible alternative. Together, the proofs for the thesis and the antithesis reveal how each side exposes inconsistencies in the alternative position. Kant describes this as 'open[ing] up a dialectical battlefield, where each party will keep the upper hand as long as it is allowed to attack, and will certainly defeat that which is compelled to conduct itself merely defensively'.[107] From this, Kant subjects both sides of the antinomy to what he terms the sceptical method. This is distinct from scepticism as the sceptical method is directed towards establishing certainty by 'seeking to discover the point of misunderstanding in disputes that are honestly intended and conducted with intelligence by both sides'.[108] In contrast, scepticism is concerned with undermining the foundation of all cognition and uprooting any possibility for certainty. Bhaskar's and Cartwright's accounts can be constructed as the thesis and antithesis of an antinomy in the following form:

> **Thesis:** Scientific research aims to discover the existence of universal transfactual laws; the laws of nature are not 'dappled'.

> **Proof:** On the assumption that there are no universal transfactual laws of nature, it follows that it would be impossible to establish transitive regularity at the empirical level in scientific conditions. We can identify regularity at the empirical level, therefore the assumption that there are no universal transfactual laws is false.

> **Antithesis:** There are no universal transfactual laws of nature; the laws of nature are 'dappled'.

> **Proof:** On the assumption that there are universal transfactual laws in nature, it follows that the regularity of such laws must be demonstrated in experience. Experience rarely reveals the manifestation of law-like regularity across nature; rather, it suggests that laws are the context-specific manifestations of regularity. These manifestations are generally the product of scientific experimental conditions. This demonstrates that the experience of laws is always context-specific, and thus the assumption of universal transfactual laws of nature is false.

Both the thesis and the antithesis agree insofar as empirical law-like regularity is not sufficient to justify the existence of universal laws. The difference is that Bhaskar argues that the partial regularity at the empirical and scientific levels requires the existence of transfactual universal laws. According to Bhaskar:

> [t]he analysis of experimental activity shows, then, that the assertion of a causal law entails the possibility of a non-human world, that it would operate even if it were unknown, just as it continues to operate when

its consequent is unrealized (or if it is unperceived or undetected by human beings), that is, outside the conditions that permits its empirical identification.[109]

Empirical scientific observations tend to oppose, rather than confirm, the regularity of transfactual laws of nature. Nevertheless, the irregularity of laws at the empirical level does not jeopardise, but in fact supports, the conception of laws as regular and universal at the non-human transfactual level. This brings into focus the differing commitments towards rationalism and empiricism held by Bhaskar and Cartwright. Bhaskar's orientation towards rationalism leads him to argue that laws 'are neither empirical statements (statements about experiences) nor statements about events. Rather they are statements about the ways of acting of independently existing and transfactually active things'.[110] In contrast, Cartwright's empiricism draws from the same irregularity of laws at the empirical level in support of the conclusion that laws are nothing more than their context-specific manifestations. Both accounts presuppose that the conditions of experience can provide access to the conditions of nature independent of experience at the non-human level. As previously noted, this does not mean that Cartwright is denying a more general conception of science as unified; she is merely denying that such unity occurs at the level of universal transfactual laws.

Her account implicitly requires the existence of universal logical laws to support her argument that laws are dappled and not universal. The non-universalisability of laws must be universal. This principle requires that both the principle of non-contradiction and the principle of sufficient reason are present throughout the universe. In other words, her account depends on the claim that laws are dappled and not otherwise, and it is possible to provide sufficient reasons why this is the case. Kant was critical of empiricism because its advocates often fail to recognise its limits and dogmatically deny the possibility of other aspects of human activities that cannot be accounted for solely in relation to appearances. According to Kant:

> if empiricism itself becomes dogmatic ... and boldly denies whatever lies beyond the sphere of its intuitive cognitions, then it itself makes the same mistake of immodesty, which is all the more blamable here, because it causes an irreparable disadvantage to the practical interests of reason.[111]

Kant's criticism of empiricism focuses on the denial of anything beyond intuitions, which follows from the extension of the conditions of knowledge specific to experience. The result is that the empiricist is unable to recognise their extension of the conditions of appearances to non-empirical domains. When applied to Cartwright's account, it is not possible to offer justification for the claim that nature independent of experience would be required to conform to the logical principles of non-contradiction and sufficient reason.

The resolution of the antinomy between Bhaskar's rationalist and Cartwright's empiricist stances towards the universal laws of nature requires an appeal to transcendental idealism; specifically, the acceptance 'that appearances in general are nothing outside our representations, which is just what we mean by their transcendental ideality'.[112] The dichotomy arises because we believe the thesis and the antithesis are the only possible explanations. As they are both exhaustive and exclusive, disproving the alternative position is sufficient proof for the correctness of the remaining side of the argument. However, Kant argues that there is a third explanation for mathematical antinomies; namely that both the thesis and the antithesis are false. According to Kant, '[i]f someone said that every body either smells good or smells not good, then there is a third possibility, namely that a body has no smell (aroma) at all, and thus both conflicting propositions can be false'.[113] In relation to this antinomy, both the thesis and the antithesis presuppose that experience is essential for identifying the conditions for the existence of universal laws. For Bhaskar, the empirical conditions that make science possible can be generalised to ontological transfactual universal laws of nature. In contrast, Cartwright argues that the majority of evidence generated by science stands in opposition to the existence of universal laws of nature. In this sense, they both extend the possibility of knowledge beyond the remit identified by Kant's critical philosophy.

Kant's philosophy can offer a middle way between antinomical positions of Bhaskar and Cartwright. Like Bhaskar, Kant also agrees that the possibility of science depends on presupposing that there are universal laws of nature. However, the need for this presupposition does not justify the ontological transfactual status of these laws. A more Kantian response is that these laws are regulative principles that are necessary for the possibility of science, but we should remain ambivalent towards the ontological status of these laws. Turning to Cartwright, our Kantian response can agree that we rarely find confirmation of the presence of universally regular laws within experience. Nonetheless, science requires us to seek unity within nature because the aim of science is to unify our experiences into a greater systematic whole. If this were not the aim of science, then science itself would be irrational as Kant argues that reason commands unity.[114]

2.3.3 Transcendental arguments in Bhaskar's and Cartwright's accounts

Both Bhaskar and Cartwright regard their philosophies as establishing transcendental certainty at an ontological level. Recall that Cartwright argues that the basis of Kant's transcendental arguments, such as the kingdom of ends or the transcendental unity of apperception, are puzzling and obscure. Instead, the basis of her transcendental argument is '[t]he objectivity of local knowledge'.[115] Stephen Clarke explains Cartwright's transcendental argument as follows: the possibility of scientific practice in general[116] requires the objectivity of local knowledge. As scientific practice in general is possible, it follows that the objectivity of local knowledge must also be possible. According to Clarke, Bhaskar's and Cartwright's transcendental arguments are incompatible:

> Each undermines the credibility of the other. Bhaskar is not merely offering us one possible way of accounting for the experimental practices of scientists. He is claiming that this is the only way to account for the experimental practices of scientists ... Cartwright's transcendental local realism involves a rival account of the elements of scientific practice that Bhaskar did not consider.[117]

Neither Bhaskar nor Cartwright successfully demonstrate what the ontology of science *must* be. There are well-established arguments against the possibility of using transcendental arguments to demonstrate certainty about the world independent of experience. According to Barry Stroud, transcendental arguments depend on the implicit assumption that, because the world appears a certain way, then it *must* be that way. Stroud argues this presupposes we are justified to make the connection between metaphysics and appearances from the fact of having that appearance. A sceptical position can show that, although we might be required to *believe* certain metaphysical principles, this does not entail the truth of those principles. According to Stroud, '[t]he skeptic distinguishes between the conditions necessary for a paradigmatic or warranted (and therefore meaningful) use of an expression or statement and the conditions under which it is true'.[118] Robert Stern summarises Stroud's position as attempting to render problematic any argument 'which asserts that "non-psychological facts" about the world outside us constitute necessary conditions for our thinking'.[119]

Both Stroud and Stern emphasise that the primary function of transcendental arguments is to defeat scepticism. They also agree that the call to answer global scepticism (i.e., doubting the existence of objects independent of experience) is beyond the remit of transcendental arguments. Transcendental arguments potentially answer the modest sceptic only if they are intended to demonstrate the necessary conditions of experience.[120] Neither Bhaskar nor Cartwright regard transcendental arguments as merely answering the modest sceptic by identifying conditions merely for the possibility of experience. In this sense, their use of transcendental arguments is problematic because both attempt to use them to ground ontological necessity. Clarke argues that neither is attempting to reject global scepticism through their use of transcendental arguments, but this must ultimately mean that they fail to demonstrate the necessity of the ontological content of their conclusions.

Conclusion

Kant's transcendental idealism was inspired in part by the limitations that he exposed in Hume's philosophy. Kant accused Hume of protecting his account of mathematics from scepticism because he had not considered that mathematics could also be derived synthetically. Kant misunderstood this aspect of Hume's philosophy as relations of ideas entail only psychological necessity that requires the demonstration of an axiom, proof or theorem as a ground for its certainty. Kant was mistaken; Hume did not distinguish between matters of facts and relations of ideas on the basis that the former related to synthetic statements and the latter to analytic statements. Rather, both were synthetic, and the difference was that knowledge of causal matters of facts could only be justified in relation to habit. By examining this misunderstanding, I laid out some of the foundations of Kant's philosophy in a way that demonstrated the similarity between Hume's philosophy and Kant's account of the synthetic *a priori*. Kant saw his own philosophy as taking the baton of philosophical issues that Hume had raised. According to Kant, if Hume had realised that relations of ideas had a synthetic ground, then this would have allowed him to revise his entire philosophy to understand how the subject could be the condition of necessity for both mathematics and causality. This helps us to understand how Kant regarded transcendental idealism as building from the foundations, and resolving the shortcomings, of Hume's empiricism.

Kant's interpretation of Hume differs significantly from Strawson's sceptical realist interpretation. The sceptical realist suggests that Hume was not an ontological anti-realist regarding causal laws, but only an epistemic antirealist. In contrast, Kant argued that Hume's philosophy had demonstrated the impossibility of any rational justification for a transcendental realist account of causal necessity. Kant's response to Hume's philosophy is foundational to transcendental idealism. He reversed Hume's irrationalism by suggesting that our capacity for reason makes it possible for us to understand the laws of nature as regulative principles. Kant's philosophy made it possible to explain causality as a condition for the possibility of experience, rather than as a relation between objects in themselves.

Kant's account causality and laws of nature offers an alternative perspective to contemporary metaphysical arguments regarding the status of laws in contemporary philosophy of science. I argued that, when formulated as a Kantian mathematical antinomy, both Bhaskar's and Cartwright's accounts are committed to variations of transcendental realism that presuppose the conditions of experience can expose metaphysical truths about the world independent of experience. They offer equally valid, yet contradictory, conclusions regarding the nature of metaphysics. The antinomy is resolved by revealing how both are committed to the problematic view that we can attain metaphysical truths from experience. The transcendental idealist solution to this antinomy combines aspects of both positions. Kant agrees with Bhaskar that the possibility of science presupposes the existence of laws. However, within the context of transcendental idealism, these laws only relate to the ability to organise experience into an architectonic system; they offer no insight into the ontological status of these laws. This also concedes Cartwright's empiricist criticism against ontological laws of nature; namely, that the majority of scientific evidence opposes the existence of universal laws. Kant's philosophy can show how the presupposition of some degree of unity is a presupposition for the existence of the possibility of science, even if empirical evidence does not confirm the existence of these laws.

3 • Kant's Influence on Whewell

Introduction

Kant's role in the development of philosophy of biology is widely recognised despite Kant's discussion of teleological judgement emerging from issues specific to the development of transcendental idealism, rather than biology. The recognition of Kant's significance for the history of biology is surprising given that Kant's account was not influential on the development of biology, at least not directly. Marjorie Grene and David Depew dedicate an entire chapter of *The Philosophy of Biology: An Episodic History* to Kant's account of teleological judgement. They argue that Kant's conception of teleology played 'only a supporting role in the overall line of argument in the *Critique of Judgment*'.[1] They under exaggerate the role of teleology in Kant's philosophy; it is central to the third *Critique*, which in turn is central to the Kant's critical philosophy. Kant explained the third *Critique* as an attempt to establish a bridge that mediates the domains of theoretical and practical philosophy through the examination of the power of judgement.[2] Kant intended his account of teleology to resolve broader tensions within transcendental idealism concerning the compatibility between the freedom and nature.

Discussions of Kant's influence on biology tend to focus solely on Kant's account of teleological judgements. For Kant, a causality other than efficient causation is required to conceive of an entity as possessing the capacity for self-organisation. This causality cannot be derived entirely from experience because it requires the subject to conceive of an entity *as if* it were produced to fulfil certain purposes, which is generated by the faculty of judgement. Kant famously argued that we could not hope for another scientist of the calibre of Newton 'who could make comprehensible even the generation of a blade of grass according to natural laws that no intention has ordered'.[3]

In this chapter I examine Kant's influence on the development of biology in the British Isles through his influence on Whewell's philosophy of science. In the first section I argue that Whewell was influenced by the architectonic structure of Kant's philosophy, in particular Whewell recognised the importance of the active powers of the mind for the acquisition of knowledge. My account offers a novel perspective on Kant's influence on Whewell. Interpreters have ranged from arguing that Kant had no influence on Whewell because of the differences between their philosophies, to arguing that Whewell is continuing the Kantian legacy. I argue that, like Kant, Whewell regarded knowledge as the product of two irreducible sources, yet Whewell was also dissatisfied with the implications that Kant's philosophy had for scientific knowledge. Whewell's account of science borrows principles from Kant's philosophy, such as the unity of science, but he extends the remit of knowledge to include things in themselves.

In the second section I examine how Whewell's philosophy of science has been influential on the development of science, yet Whewell's theological justification for his account of science stands in tension with more recent appeals to consilience. Specifically, I focus on Edward Wilson's account of consilience and the assumptions of consilience behind accounts such as John Mackie's argument against objective morality. These accounts of consilience offer naturalist accounts that place minimal importance on the role of God, yet historically consilience had the impact on science because of the role that God played for scientific knowledge. I argue that Whewell escaped the implications of transcendentalism because he justified scientific knowledge by appealing to God. The comparison between the accounts of Kant and Whewell exposes the metaphysical cost attached to some assumptions that are central aspects of the general understanding of science. Kant's account of science offers an alternative account, which avoids this metaphysical cost by developing a regulative account of scientific unity.

3.1 The relationship between Whewell and Kant

Kant's influence on Whewell has been the source of dispute in scholarship as there are significant differences between their philosophies. I argue that Kant's influence is evident in the following three aspects of Whewell's philosophy: first, his account of the active powers of the mind; second, his account of knowledge as the product of two irreducibly distinct sources; and third, his account of consilience. Whewell was dissatisfied with various aspects of Kant's critical philosophy, including his reduction of possible knowledge to the fixed forms of the categories and his denial of knowledge of things in themselves. For Whewell, these aspects of Kant's philosophy had significant limitations for scientific knowledge. Against Kant, Whewell argued that scientific knowledge related to things in themselves. Knowledge is the product of the fundamental antithesis between thoughts (or ideas) and things. New scientific theories arise when ideas are applied to things in a novel way. I outline the different stages of Whewell's account of scientific knowledge, such as the colligation of facts and the consilience of inductions. Whewell's commitment to realism, in contrast to Kant's transcendental idealism, requires Whewell to demonstrate how consilience is not merely a pragmatic virtue for knowledge, but also is an inherent feature of nature independent of experience. Finally, I examine the significance of God for the accounts from Kant and Whewell. For Whewell, the good fit between our explanation and nature is evidence for God's existence. In contrast, Kant denied that we could formulate any possible knowledge of the existence of God.

3.1.1 The similarities and differences between the philosophies of Kant and Whewell

Kant's influence on Whewell is widely (but not unanimously) accepted in the literature. According to Steffen Ducheyne, 'Whewell learned from Kant's philosophy the importance of the active powers of reason ... He stressed that in order to know we must perceive and conceive. Knowledge implies both passive as well as active thought'.[4] Whewell's account differed from transcendental idealism because he regarded the categories and intuitions as both belonging to 'ideas'. According to Robert Butts, 'Whewell's Fundamental Ideas are simply Kant's forms of intuition and categories under a new name'.[5]

Like Kant, Whewell regarded space, time and causality as conditions of experience that were derived from the subject. However, the argument that categories and intuitions both belonged to the same faculty was specific to Whewell. Recall that, for Kant, intuitions refer to the way that objects are given to us in space and time, whereas concepts relate to the way that objects are thought.

Only the former guarantees the existence of its object as it is possible to think of an object without any such object existing, but the object is *given* in experience through intuition.[6] These faculties are different in kind because each has different functions. According to Kant, 'these two faculties cannot exchange their functions. The understanding is not capable of intuiting anything, and the senses are not capable of thinking anything. Only from their unification can cognition arise'.[7] Sensibility furnishes the understanding with the content of knowledge; the understanding 'can never overstep the limits of sensibility, within which alone objects are given to us'.[8] This does not mean that the understanding literally *cannot* overstep the limits of sensibility, rather the understanding *should* not overstep the limits of sensibility because the categories can only generate knowledge when they are applied to appearances. The categories could only apply to things in themselves if the conditions for things in themselves and appearances are identical. For Kant, this is fundamental to the distinction between transcendental idealism and transcendental realism.[9]

When considered from the perspective of transcendental idealism, Whewell does not fit easily into transcendental idealism or transcendental realism. Whewell is not a transcendental realist insofar as he does not regard the conditions of things in themselves as identical to the conditions of experience. According to Whewell, knowledge of necessity cannot be derived from experience: 'Necessary truths derive their necessity from the ideas which they involve; and the existence of necessary truths proves the existence of Ideas not generated by experience'.[10]

Whewell's argument that ideas are generated by a source other than experience shows how aspects of Whewell's philosophy align with transcendental idealism. However, there are important differences between their accounts, which are indicative of the broader tensions between their accounts. For instance, the role of necessity differs significantly between their accounts. It is helpful elaborate on this by examining their alternative responses to Hume.

Kant accepted Hume's argument that causal necessity was deceptive and illusory if it related to objects in themselves. Kant's separation of objects in themselves from objects of appearance enabled him 'not only to prove the objective reality of the concept of a cause with respect to objects of experience but also to *deduce* it as an *a priori* concept'.[11] Whewell agreed that causal necessity cannot be experienced as a property of objects in themselves. How-ever, he denied Kant's solution that we must recourse to objects of appearance at the cost of knowledge of things in themselves to discover this source of necessity. Thus, Whewell rejected Humean scepticism: '[o]ur inference from Hume's observation is, not the truth of his conclusion, but the falsehood of his premises; – not that, therefore, we can know nothing of natural connexion, but that, therefore, we have some other source of knowledge than experience.'[12] Hume's argument that the source of necessity could not be derived from experience of objects led Whewell to argue that it must be derived from a different source; namely, ideas. While he acknowledged the difference (the fundamental antithesis) between things and ideas, he did not regard this as entailing that experience and knowledge are not directly related to objects in themselves. According to Whewell, 'We have an Intuition of objects in space; ... and apprehend their spatial relations by the same act by which we apprehend the objects themselves'.[13] His claim that by appre-hending the object in intuition we are also apprehending the object in itself marks a significant departure from Kant's conception of transcendental idealism. Kant denied our apprehension of things in themselves precisely because space and time are conditions of experience alone. He argues for 'the empirical reality of space (with respect to all possible outer experience), though to be sure at the same time its transcendental ideality, i.e., that it is nothing as soon as we leave out the condition of the possibility of all experience, and take it as something that grounds the things in themselves'.[14]

Kant distinguished between concepts and intuitions as a dif-ference in kind, whereas Whewell regarded both intuitions and concepts as belonging on the side of ideas. This may have been because Whewell thought that both were fundamentally active features of the mind within Kant's philosophy. Kant described the faculty of sensibility as a passive or receptive faculty in contrast to the faculty of understanding, which is active or spontaneous. How-ever, it is a mistake to make this distinction between activity and passivity the essential difference between these faculties. According to Watkins:

> sensibility is still active insofar as it does not literally *receive* repre-sentations from without, but rather *actively* forms representations ... if sensibility were essentially characterized *solely* in causal terms, then

both sensibility and understanding would be active and would differ only by means of degree, contrary to Kant's position.[15]

Whewell does consider sensibility and understanding as differing in degree, and in this sense his account is contrary to Kant's for the reason outlined by Watkins. There is an underlying similarity between their philosophies as both regarded knowledge as the product of two irreducible elements. In contrast to Kant, who argued that the foundation of knowledge was intuitions and concepts, Whewell argued that the foundation of knowledge was thoughts and things. Whewell's identification of both categories and intuitions with ideas (or thoughts) enabled him to locate the other irreducible source of knowledge in things. According to Whewell, '[i]n all cases, Knowledge implies a combination of Thoughts and Things. Without this combination, it would not be Knowledge. Without Thoughts, there could be no connexion; without Things, there could be no reality'.[16] Whewell termed the relationship between thoughts and things the fundamental antithesis and described it as 'the ultimate problem of all philosophy'.[17] The structural similarity of this passage with a well-known passage in Kant's first *Critique* is evidence to suggest that Whewell regarded the relationship between thoughts and things as analogous to the relationship between intuitions and concepts within Kant's philosophy. According to Kant, '[w]ithout sensibility no object would be given to us, and without understanding none would be thought'.[18]

The problem that both Kant and Whewell faced was explaining how knowledge was the product of two irreducibly distinct sources. Whewell commented on the strange relationship that thoughts and things have with one another; from the perspective of knowledge they must be considered as united, but from the perspective of philosophy they must be considered as separate.[19] Butts describes the relation between thoughts and things as susceptible to a version of the third man argument.[20] According to Butts, 'there must be some principle of unity that subsumes both terms in the pairs that Whewell lists. There must be some "third man" (the problem is a variation on Kant's schematism problem)[21] which is like both Ideas and Things'.[22]

The schematism is Kant's attempt to demonstrate the transcendental necessity of experience as the combination of both concepts and intuitions, while simultaneously conceding that knowledge of how this happens is beyond the discursive limits of cognition. It is one of the most obscure and difficult aspects of the first *Critique*.[23] Kant admits that the schematism is shrouded in mystery as he describes it as 'a hidden art in the depths of the human soul'.[24] To cast some clarity on the schematism, it is helpful to consider its function within Kant's critical philosophy. A crucial aspect of transcendental idealism is its ability to demonstrate that the origin of experience emerges neither from some form of intellectual intuition (which he argued was a prevalent view among rationalist philosophers), nor from the direct experience of objects in themselves (which Kant attributed to empiricism). In the context of transcendental idealism, the schematism reinforced Kant's attack on rationalist metaphysics by appealing to the importance of both intuitions and concepts for possible experience. According to Walsh, 'Kant's case against traditional metaphysicians ... is that having only the pure categories on which to build they succeed neither in saying anything precise nor in saying anything about anything in particular'.[25] Kant emphasises the importance of applying our categories to objects in sensible intuition, as categories without intuitions are empty.[26]

The schematism simultaneously displays both Kant's ambitious rejection of transcendental realist metaphysics and the underlying humility of transcendental idealism. This humility consists in the idea that there is a limit to knowledge, experience and explanation. This is not, as Rae Langton suggests, 'a depressing discovery' because 'Kant thinks we are missing out on something in not knowing things as they are in themselves'.[27] It is difficult to understand how we could be missing out on something that we are not justified to posit knowledge of in the first place. Instead, Kant's philosophy demonstrates how transcendental idealism secures the objectivity of knowledge precisely because it limits theoretical knowledge to experience. For instance, in his response to the argument that the ideality of space and time reduces experience to mere illusions, Kant emphasises that knowledge of the *a priori* of mathematics and geometry is only possible by understanding objects as appearances. He argues that if space and time lacked *a priori* foundations, then we could not rule out that our intuitions, or even geometry, are mere 'self-produced brain phantoms' or 'illusions', which do not correspond to any objects. Hence, Kant turns back the charge of experience as illusory against those who deny that space and time are conditions of experience rather than properties of objects in themselves. If space and time are properties of objects, then we cannot know that those properties will remain constant for all experience across time. According to Kant, 'we have been able to demonstrate the incontestable validity of geometry with respect to all objects of the sensible world for the very reason that the latter are mere appearances'.[28]

The role of the schematism is to explain how experience is a combination of intuitions and concepts, while simultaneously arguing that these are two irreducible faculties. Kant describes the schema as 'the phenomenon, or the sensible concept of an object, in agreement with the category'.[29] This provides additional support for Kant's rejection of space and time as properties of objects in themselves because it emphasises that objects are generated by the subject.[30] Although the need for the schematism is clear for Kant's critical philosophy, the specific detail about how it unifies concepts and intuitions into experience is far from straight forward. According to Gardner, '[t]he obscurity attaching to the doctrine of schematism is the price which Kant ultimately pays for escaping from rationalism and empiricism, and rejecting the transcendental realist model of concept application'.[31]

Despite the obscurity of the schematism, understanding its function within transcendental idealism highlights both the similarities and differences between Kant and Whewell. Whewell does not share Kant's opposition to transcendental realism in relation to metaphysics, nor does he deem it necessary to resolve the fundamental antithesis between thoughts and things by appealing to something analogous to Kant's schematism. Instead, Whewell proposed a theological resolution to the fundamental antithesis. Despite this point of divergence, a fundamental structural similarity remains between their accounts. Central to both is the claim that knowledge is the product of two irreducible elements. It seems likely that this aspect of Whewell's conception of the fundamental antithesis was a partial response to Kant's account of knowledge as comprised of irreducible components of intuitions and concepts, and his divergence was, in part, due to his dissatisfaction with Kant's account of science.

Kant could not develop an appropriate philosophy of science for Whewell because of his denial of knowledge of things in themselves. According to Whewell, Kant made things in themselves 'a

dim and unknown region. Things were acknowledged to *be* something in themselves, but *what*, the philosopher could not tell'.[32] The
implication of critical idealism is that things in themselves can only
be thought, yet never known. These entities necessarily remained
beyond the domain of experience: 'appearances are ... given not
in themselves, but only in this experience, because they are mere
representations, which signify a real object only as perceptions.'[33]

Whewell's dissatisfaction with Kant's philosophy is further illus-
trated by his discussion of Kant's comparison of transcendental
idealism with the Copernican revolution. The first *Critique* begins
by proposing a transformation in our way of thinking that is at
odds with sensory experience, but it is then demonstrated with
apodictic certainty. Kant praised the Copernican revolution as an
example of such a transformation.[34] Whewell argued that Kant's
philosophy was in tension with the Copernican revolution as it
had not sufficiently accounted for the conclusions drawn from it:

> Kant conceives that our experience is regulated by our own faculties,
> as the phenomena of the heavens are regulated by our own motions ...
> we may say that Kant, in explaining the phenomena of the heavens by
> means of the motions of the earth, has almost forgotten that the planets
> have their own proper motions, and has given us a system which hardly
> explains anything beside the broadest appearances.[35]

Kant had not merely forgotten that the planets have their own
proper motions; critical idealism entailed that we must regard our
knowledge of these motions as regulative (or reflective), rather
than constitutive (or determinative). This distinction refers to
the two possible forms of judgement in Kant's philosophy: '[t]he
power of judgment is of two kinds: the *determinative* or the *reflective* power of judgment. The former goes *from the universal to the
particular*, the second *from the particular to the universal*. The latter has only *subjective* validity.'[36] Determinative judgements relate
to the twelve pure categories that are necessary for the possibility
of experience. They 'provide the appearances with their lawfulness
and by that very means make experience possible'.[37] The primary
function of the faculty of understanding is not to establish scientific laws, which build on the content of experience. Rather the
pure categories are rules that are necessary for the possibility of
experience. Kant describes them as an exhaustive 'listing of all

original pure concepts of synthesis that the understanding contains in itself *a priori*'.[38]

The dissimilarity between the Kantian and Copernican revolutions is that Kant regarded science is an enquiry into the systematising capacity of the subject to project an order onto nature, whereas Copernicus had sought to discover the laws of nature that persist independently of the subject. In a similar line of argument, Quintin Meillassoux charged Kant's analogy to the Copernican revolution with committing a violent contradiction. Kant's reorientation of scientific knowledge from a realism pertaining to truths that are independent of experience to an idealism that re-established the subject at the centre of knowledge, is returning to an essentially Ptolemaic conception of science. According to Meillassoux, Kant's philosophy is more appropriately described as the Ptolemaic counter-revolution as 'it is only since philosophy has attempted to think rigorously the revolution in the realm of knowledge brought about by the advent of modern science that philosophy has renounced the very thing that constituted the essence of this revolution'.[39] The essence of the revolution of modern science was its ability to go beyond phenomena by utilising mathematics to establish knowledge about objects independent of experience.

While Whewell does not explicitly identify Kant's philosophy with the Ptolemaic scheme, his description of the Ptolemaic scheme is strikingly similar to his description of Kant. He describes the Ptolemaic scheme as 'the view of those who appeal to phenomena alone as the source of our knowledge, and say that the sun, the moon, and the planets move as we see them move, and that all further theory is imaginary and fantastical'.[40] Central to both is the appeal to phenomena as the source of knowledge and the inability to extend the same level of certainty for the celestial motions of the objects in themselves.[41]

The similarity between Whewell's description of the Ptolemaic and Kantian account of science potentially reveals reasons why Whewell avoided committing himself to Kant's transcendental framework. In his view, Kant's philosophy was implicitly committed to the renunciation of the scientific developments that it had appealed to in support of the revolutionary shift towards transcendental idealism. Kant did not renounce the developments in science, but rather he attempted to relocate our understanding of scientific truths away from the idea that they pertain to things in themselves.[42]

For Kant, knowledge of scientific laws required that an empirical concept corresponding to an intuition is given *a priori*.[43] In the *Metaphysical Foundations of Natural Science*, Kant identified the empirical concept of matter as a case that achieves this because the experience of matter is inseparable from the *a priori* intuitions of space and time. This sufficiently demonstrates the conditions for the empirical concept of matter as a particular law for every possible experience of outer sense. According to James Kreines, the empirical concept of matter is a special case where synthetic *a priori* necessity is achieved; however, more generally, 'we can clearly see that Kant is right to rule out the possibility of similar derivations for particular laws'.[44] In the context of his critical philosophy, Kant is correct to argue that our inability to establish *a priori* principles for particular scientific laws is not a result of deficient knowledge at any point in time. Rather, 'the limit is *in principle* and ineliminable; our lack of knowledge stems not from the state of science at some particular time but from our limited access to *a priori* intuition, or from "the limits of our faculties of cognition"'.[45]

In contrast, Whewell argued that objects in themselves possess the power to *inform* ideas related to a scientific understanding of nature. Thus, he opposed Kant's view that pure intuitions and pure categories formed an exhaustive account of the possible conditions of knowledge. According to Menachem Fisch, Whewell 'rejected Kant's claim that the list of the two forms of intuition and the categories discernible through their role in former knowledge, exhausts the range of possible categories of any future knowledge'.[46] Fisch regards Whewell's rejection of these aspects of Kant's philosophy as evidence of the lack of influence of Kant on the development of Whewell's philosophy of science. Fisch argues that Whewell did not reference Kant in his published or unpublished works during the early 1830s, when Whewell was developing his philosophy and epistemology of science. There were 'no questions of epistemology, of the possible boundaries of human knowledge ... no categories, no forms of intuition, no talk of analytic versus synthetic *a priori* truths'.[47] In opposition to the idea that Kant was not influential on Whewell, Ducheyne has identified various references in Whewell's unpublished notebooks (between approximately 1830–3) that provide evidence for Kant's influence on Whewell. For instance, Whewell described space and time as intuitions and explicitly appealed to Kant's term '*Anschauung*'.[48]

Another reason why Whewell might not have adopted Kantian terminology in his published works is due to the negativity towards German philosophy in the British Isles at this time. Andrew Cooper explains how an anonymous contributor to *The Anti-Jacobian Review and Magazine* personally attacked Coleridge and Wordsworth after they returned from Germany and expressed sympathy for German ideas. According to Cooper, '[t]he author presents Coleridge's passage to Germany as a matter of national betrayal'.[49] It is probable that the hostility towards German ideas within the British Isles contributed to Whewell's lack of reference to Kant in his published works.[50] Whewell's lack of reference to Kant during the development of his philosophy of science is more likely due to several factors. Clearly, there are important differences between the two philosophers, but these reflect Whewell's dissatisfaction with transcendental idealism. Whewell did not engage in the problems specific to Kant's critical philosophy because Kant denied knowledge of things in themselves and relegated them to an unknown X.[51]

3.1.2 Whewell's colligation of facts and the consilience of inductions

Whewell's antithesis between thoughts and things enabled his philosophy of science to develop in a different direction than was possible in accordance with Kant's transcendental idealism. Whewell asserted that knowledge is in a direct relationship with objects in themselves. His account of science built from the foundations of the antithesis between thoughts and things to develop another level of antithesis between facts and theories. According to Whewell, 'a Fact is a combination of our thoughts and things in so complete agreement that we do not regard them as separate'.[52] In contrast, a theory is distinct from facts, yet dependent on them. A theory is an idea that goes beyond what is currently known as fact. A theory has the potential to become true knowledge if it can effectively colligate multiple facts. According to Butts, '[t]o get science, we need ideas that are clear and distinct, and that adequately colligate facts in such fashion that general propositions about matters of fact become possible'.[53] Whewell describes the colligation of facts as the process by which scientists form testable hypotheses that have the potential to combine 'scattered facts into a single rule'.[54] This does not mean that every hypothesis will adequately colligate facts, rather it merely entails that the colligation of facts is a requirement for science. The scientists' ability to produce clear and distinct ideas that bind facts together under a general rule cannot be taught in accordance with any pre-existing rule. According to Whewell, 'it more frequently happens that new truths are brought into view by the application of new Ideas, not by new modifications of old ones'.[55] Hence, this capacity belongs to the sagacity or wisdom of particular scientists.

The colligation of facts is similar to Kant's account of the regulative unity of science in some important respects. For both, establishing unity is not merely an enumeration of examples, as this could not justify the necessity that such principles hold in all cases. In contrast to Kant, the colligation of facts does not correspond to regulative principles of reason, rather it serves as evidence for the truth of a theory. For any successful colligation of facts, a clear and distinct idea is superinduced via means of a hypothesis not merely on the current facts belonging to a class, but any future phenomena belonging to that same class. Hence, a colligation of facts should accurately predict future facts of the same class where the hypothesis would apply. According to Whewell, '[t]he prediction of results, even of the same kind as those which have been observed, in new cases, is a proof of real success in our inductive processes'.[56]

The greatest evidential force of the correctness of a theory is achieved when a hypothesis can be applied to cases involving a class of facts that differs from the original class that the hypothesis explained. According to Whewell, '[t]hat rules springing from remote and unconnected quarters should thus leap to the same point, can only arise from *that* being the point where truth resides'.[57] Whewell terms this 'jumping together' of facts belonging to different classes under a single rule 'the consilience of inductions'.

Whewell's justification for the principle of consilience is both historical and philosophical. In the historical context, he argues that there has been no occasion in the history of science where a theory has achieved consilience and has later been proven false.[58] One example that Whewell discusses is the discovery of universal gravitation. Not only did it explain the perturbations of planets in relation to each other and to the sun, but it also explained axial precession. Whewell describes the ability for the theory to explain this coincidence as giving the theory 'a stamp of truth'.

Whewell's Kantian influence has been emphasised by Butts: 'Whewellian consilience is the prodigy of Kantian systematization'.[59] While there are significant similarities between Kant and Whewell in relation to their accounts of the systematisation of nature, there are also equally significant differences. Recall that for Kant, most scientific laws are denied the possibility of being known with *a priori* certainty. Despite lacking *a priori* certain knowledge of laws, we can effectively appeal to laws as regulative principles that guide scientific enquiry. Kant describes how the principle of gravitation is used to explain the motions of the planets and comets. Such principles can be followed:

> merely by approximation, without ever reaching them, yet these principles, as synthetic propositions *a priori*, nevertheless have objective but indeterminate validity, and serve as a rule of possible experience, and can even be used with good success, as heuristic principles, in actually elaborating it; and yet one cannot bring about a transcendental deduction of them.[60]

A transcendental deduction is not possible for these principles because they are not necessary conditions for the possibility of experience, but only necessary conditions for our ability to engage in scientific activity. They are heuristic principles, but this does not entail that their deployment is anything less than law governed. These heuristic principles impose a systematic unity and completeness on nature that cannot be derived from the faculty of sensibility (i.e., intuitions) or the faculty of the understanding (i.e., concepts). Rather, the faculty of reason subsumes the relevant experiences under a general rule. Reason projects a complete unity and order onto nature. According to Kant, this 'completeness can ... only be understood as a completeness of principles, but not of intuitions and objects'.[61] Therefore, it is not possible to regard this unity as derived from our experience of nature, nor does it pertain to any object independent of experience.

The connotations attached to the term 'heuristic' are misleading as a description of these regulative principles. Allison elaborates on the misleading connotations associated with the term heuristic by drawing on Kant's dictum that 'reason does not beg but commands'[62] unity. According to Allison, 'to deny that reason begs is to deny this principle is a merely heuristic device through which it approaches nature, hat in hand, as it were, hoping to find some degree of confirmation'.[63] To be clear, regulative principles are similar to heuristic principles in the sense that they are deployed by the subject. However, unlike heuristic principles, regulative principles are neither a matter of choice nor chance, but a precondition for the possibility of scientific enquiry.

Rachel Zuckert helps us to appreciate how Kant argues both that we can presuppose something and simultaneously deny sufficient evidence to claim that it exists. She argues that the ideas motivating scientific enquiry do not always refer to objects. She describes this relationship of 'ideas without objects' as acting as placeholders for scientific development. According to Zuckert, 'we need to have a sense, a projected image of such a world, as something "out there" that might be investigated (profitably). We need (what I call) an optimistic placeholder, a stand-in for the "I know not what" we hope to find in investigation'.[64] These placeholders are sufficient to motivate us towards further scientific enquiry, but do not guarantee that we will get to a stage where we gain sufficient evidence to demonstrate that these placeholders refer to existing objects. They are vague enough to require us only to think that science could offer better explanations of certain phenomena in the future, without demanding that we know how this will be achieved.

This marks a significant difference between the philosophies of Kant and Whewell. In a sense, Kant was concerned with demonstrating the boundaries of knowledge, which has significant limitations for scientific knowledge. In contrast, Whewell was not concerned with restricting the domain of scientific knowledge, instead he intended to develop a philosophical methodology that was appropriate *for* science.[65] He was motivated to supplement the scientific methodology of his time with a philosophy that supported the development of science. He recognised the limitations of transcendental idealism insofar as it denied theoretical knowledge of objects in themselves and reduced the domain of knowledge to intuitions and concepts. Within Kant's philosophy, the idea that science was progressing towards a more unified explanation could only be explained as a regulative demand of reason.

As previously suggested, one reason why Whewell may not have adopted the terminology specific to transcendental idealism is because he wanted to avoid committing himself to the limitations he exposed within Kant's philosophy. This does not mean that Kant did not influence Whewell. It is clear he was influenced by three elements of Kant's philosophy: first, his emphasis on the active powers of the mind; second, his claim that knowledge is formed from two irreducible elements; and third, his demand for a unified science. Their fundamental difference is that where Kant sought to develop a philosophy that could both guide and limit scientific enquiry, Whewell reversed the relationship between philosophy and science. He argued that philosophy should be guided by the wisdom and sagacity of scientists in their discovery of new theories and their ability to test objects in themselves for potential confirmation of these theories. According to Whewell:

> Not only do I hold that the Axioms, on which the truths of science rest, grow from guesses into Axioms in various ways, and often gradually, and at different periods in different minds … but I conceive that this may be shown by the history of science, as having really happened, with regard to all the most conspicuous of such principles.[66]

Whewell's appeal to history in support of his account of scientific truth is indicative of many underlying issues for his theory. From the perspective of transcendental idealism, Whewell's approach at best only secures the fact that his proposed account of science is compatible with the history of science. This cannot offer sufficient justification for Whewell's assertion that he has discovered the procedure for scientific theories to establish the correct explanations, which culminate in a progressive and unified science.

The reason that history cannot offer sufficient justification to science is similar to Kant's argument against Locke's physical derivation of the self.[67] Recall that Kant argued that a physiological explanation lacked the capacity to identify the *a priori* properties of the self; 'in regard to pure *a priori* concepts empirical deductions are nothing but idle attempts, which can occupy only those who have not grasped the entirely distinctive nature of these cognitions'.[68] A historical justification of scientific truth can only offer justification at the empirical level; it cannot reveal the *a priori* conditions of scientific truth. Whewell gets around this problem by rejecting the notion that there is a difference in kind between *a posteriori* and *a priori* truths. Instead, there is only a difference of degree rather than kind; '*a posteriori* truths become *a priori* truths'.[69]

3.1.3 Whewell's theological resolution of the fundamental antithesis

Whewell appeals to the role of God to support his account of science, in contrast Kant denied any possible knowledge of the existence of God in his refutation of the ontological argument. There is an intricate relationship between science and theology within Whewell's philosophy as he appeals to theological principles to resolve many issues within his account. These aspects of Whewell's philosophy go far beyond the limits of knowledge specified by transcendental idealism. I argue that Kant's refutation of the ontological argument is significant not merely as a refutation of Whewell's supporting argument for his belief in the existence of God, but as a refutation of Whewell's justification for his philosophy of science.

The prevalent explanation for Whewell's resolution of the fundamental antithesis is his appeal to theology. According to Whewell, 'our Ideas correspond to the Facts of the world, and the Facts to our Ideas, because our ideas are given us by the same power which made the world, and given so that these can and must agree with the world so made'.[70] Our scientific knowledge is not equivalent to God's knowledge. Whewell appeals to the humility of our own scientific knowledge in comparison to the ideas in the divine mind. God 'sees the essence of things through all time and through all space; while we, slowly and painfully, by observation and experiment … make out a few of the properties of a kind of thing'.[71]

Whewell's appeal to theology allows him to justify his belief that science is our way to understand things in themselves. In other words, it allows us to understand things from a God's eye view, yet we only see a glimpse of the essence of things as God sees them. For Whewell, the connection between science and theology makes it possible for us to understand the necessary properties of objects, because we have a partial understanding of the essence of things. Whewell denies that we could arrive at complete knowledge of the world as the divine mind conceives it. He appeals to metaphors of the ocean to elucidate the difference in the magnitude of divine knowledge in comparison to our own. He describes our science as a drop in the ocean of what is known to the divine mind, and our science extends only so far as it keeps its footing 'in the shallow waters which lie on the shore of the vast ocean of unfathomable truth'.[72]

Whewell's appeal to theology as justification for our scientific knowledge reveals the dual aspects of extravagance and humility within Whewell's philosophy, which are inverted in Kant's philosophy. Whewell's account is extravagant insofar as the claim that necessity can be inferred from the necessity of the divine mind is essentially the dismissal of Hume's problem of induction. Whewell learnt from Hume's accounts that necessity must be derived from a source other than experience. His conclusion differed significantly from Hume's conclusion that there is no rational argument for our belief in causality understood as a necessary connection between objects. Like Hume, Whewell agreed that necessity could not be derived from experience, but he saved knowledge of necessity by making it dependent on the existence of God. Alternatively, Kant limited causal necessity to a condition of the possibility of experience. This entailed the denial of the possibility of any knowledge of causal necessity about objects independent of our experience. According to Kant, '[t]he non-sensible cause of these representations is entirely unknown to us, and therefore we cannot intuit it as an object'.[73] Hence, Kant rejected any possibility that empirical knowledge could offer any insight into objects in themselves, as both the properties and origin of such objects are entirely unknown to us.

In relation to knowledge of the existence of God, Kant famously exposed the problematic status of the ontological proof of God's existence. He argued that being 'is obviously not a real predicate, i.e., a concept of something that could add to the concept of a thing'.[74] The statement 'God exists' can be understood as either an analytic or synthetic statement. Kant argues that if it is meant analytically, then it is tautological because the predicate of existence would be contained in the subject of God. In contrast, if the statement is meant synthetically, then it would be impossible to conceive of the possibility that the predicate could be false without contradiction. The existence of God would already be demonstrated if God was an empirical object. According to Kant, '[i]f the issue were an object of sense, then I could not confuse the existence of a thing with the mere concept of a thing'.[75] The appropriate domain for the deployment of concepts is always empirical. In cases where there is no corresponding experience and we attempt to 'think existence through the pure category alone ... it is no wonder that we cannot assign any mark distinguishing it from mere possibility'.[76] Hence, Whewell's argument that God has complete knowledge of ideas, whereas we only have a limited knowledge, is not justified from the perspective of transcendental idealism. Whewell's argument requires a demonstration of the existence of God, which Kant argues is beyond the possibility of knowledge. Hence, Kant would oppose Whewell on the basis that reason, not God, is the source of our ideas. According to Kant:

> [A]ll the questions that pure reason lays before us, lie not in experi-
> ence but themselves in turn only in reason, and they must therefore be
> able to be solved ... since reason has given birth to these ideas from its
> womb alone, and is therefore liable to give an account of either their
> validity or their dialectical illusion.[77]

Kant denies the idea that science provides a God's eye view of the world. It is illegitimate to regard our own reason as a deficient form of God's reason because we cannot know that God exists. Kant argues that God is a regulative principle. According to Kant, 'reason's supposition of a highest being as the supreme cause is thought merely relatively, on behalf of the systematic world of sense, and it is a mere Something in the idea, of which we have no concept of what it is in itself'.[78]

Whewell's need to appeal to God to resolve the fundamental antithesis – which is central to his justification for his philosophy of science – is significant because it reveals his need to ground the unity of nature in something other than our own rational cap-acity. Not only is this attempt to justify his philosophy of science theologically insufficient from the perspective of transcendental idealism, but it also has important implications for both contem-porary interpretations of Kant's philosophy that attempt to make Kant's philosophy compatible with (or supportive for) contem-porary philosophy of science and contemporary philosophies of science that continue to appeal to consilience.

3.2 The status of consilience in contemporary philosophy of biology

In this section, I examine the importance of the unity of science for interpretations of Kant's philosophy and for philosophy of science. First, I consider contemporary interpretations of Kant that suggest he can offer support to philosophy of science. The accounts that I discuss are Philip Kitcher's appeal to the historical inheritance of scientific theories and Angela Breitenbach's and Yoon Choi's stance of unified pluralism. I argue that neither account adequately appreciates how Kant's regulative account of scientific laws differs from the metaphysical commitments of contemporary philosophy of science. Second, I examine recent appeals to consilience in philosophy of science. I consider the role of consilience in arguments for biological reductivism from John Mackie and Edward Wilson. Both place a greater naturalist metaphysical demand on consilience than Whewell's original formulation. Whewell's account of consilience was not a universal demand on nature but rather a context-specific proof of a theory that was justified by appealing to God. In contrast, recent appeals argue for the metaphysically reductive account of the universal principle of consilience, which aims at developing an all-encompassing scientific explanation of reality.

3.2.1 Contemporary interpretations of Kant's philosophy of science

Interpretations of Kant tend to focus on the extent to which his philosophy can offer support to contemporary philosophy of science, rather than examining the incompatibilities between their accounts. Kitcher proposes a broadly Kantian account of science that appeals to the historical indebtedness of scientific theories. He aims to purge Kant's philosophy of its commitment to *a priori* elements. There are many aspects of Kant's philosophy where *a priori* conditions are offered as the basis for justification, so it is important to understand precisely which aspects of Kant's *a priority* that Kitcher's account is rejecting. He opposes the demand for *a priori* certainty that is central to Kant's distinction between proper and improper sciences on the basis that the former 'treats its object wholly in accordance to *a priori* principles'.[79] Kitcher emphasises that the regulative status of scientific enquiry seems to be at odds with the stringent demand that Kant places on science. Instead, he argues that we should look to the sciences that we have inherited, science is minimally 'choosing the better over the worse, or, in the actual conduct of science, selecting the best of the options that have been explicitly formulated'.[80] Science is inseparable from history as we inherit our ideas of science from our predecessors. According to Kitcher:

We do not go through any explicit process of systematizing our beliefs and attempting to maximize the unity of the system. We absorb from our predecessors the order of nature that they have projected, so that, from the beginning of our own discussions of the world of experience, we tacitly operate with claims about causal dependencies and natural kinds that have been generated by the systems of our ancestors. Our justifications are thus parasitic on the history of attempts to construct a systematic unification of human experience.[81]

Kitcher does consider that many of the previous scientific theories are not compatible with the conditions imposed on knowledge by transcendental idealism. Most scientists would deny that their scientific theories are only projections of order onto nature. They would tend to regard their theories as metaphysical explanations of the fundamental elements of reality. If it is possible to appeal to Kant's philosophy to explain how we inherit these systematic accounts of nature from our predecessors, then we must also determine whether these theories are compatible with the broader account of scientific knowledge developed by Kant.

Kant described the first *Critique* as a ground clearing exercise[82] because his philosophy identified the limits of knowledge and highlighted 'that human reason has the propensity to overstep all these boundaries'.[83] The unity of nature is one example of how reason can overstep the boundaries of knowledge when considered as a metaphysical principle. Kant aimed to demonstrate that we can use the unity of nature as a regulative demand of reason to guide scientific enquiry in a way that does not overstep the boundaries of reason.

According to Allison, Kant's appeal to the example of the *focus imaginarius* reveals how Kant regarded the unity of nature as indispensable yet ultimately fictitious.[84] The *focus imaginarius* relates to an optical illusion whereby a subject views an image in a mirror and perceives an object physically behind the mirror. In its empirical context, the illusion is optical, but in this context, the illusion is transcendental. For both cases, the illusion does not reside in its presentation, but from what is inferred from that presentation. For Kant, we are justified to view nature as unified insofar as we recognise the role this hope plays in motivating us to seek greater unity within our scientific understanding. This does not entail that science must inevitably reach a point where it will be possible to exhaustively identify objects and laws that are relevant to science.

The problem with Kitcher's account is that many scientific theories make claims that extend beyond the limits of knowledge specified by transcendental idealism. Transcendental idealism would reject any scientific theory that makes knowledge claims about objects in themselves. Such theories would be committed to a version of transcendental realism, which is incompatible with transcendental idealism. In addition, if multiple theories were compatible with transcendental idealism, it would be impossible to assess which of these theories was correct. These theories would be compatible with Kant's broader account of scientific unity as an indispensable and indeterminate product of reason that cannot correspond to, or be sufficiently demonstrated by, any empirical object. These theories would develop in accordance with the regulative use of reason, which cannot be in antithesis or contradiction with itself.[85] In this circumstance, Kant's philosophy could only support science by outlining the conditions for science by 'determining and judging what is lawful in reason in general'.[86]

Kitcher's account of judging the best scientific theory by selecting from available alternatives is closer to Whewell's account. Whewell appeals to the history of science as justification for his account of the consilience of inductions. Those theories that have manifested consilience in the past have never been proven false. These appeals to the history of science stand in tension with Kant's aim to outline the conditions for how science can develop in accordance with reason. Instead, Whewell and Kitcher focus on identifying the aspects of science that have worked historically and suggest that these aspects are the most appropriate foundation for us to model contemporary science on.

Another account that has suggested philosophy of science could benefit from adopting a Kantian spirited methodology is Breitenbach's and Choi's account of 'unified pluralism'. In contrast to Kitcher's argument that focuses on the historical development of science, they consider the growing support for pluralism in philosophy of science. Pluralism is generally opposed to the unity of science perspective, as pluralist approaches simultaneously deploy many different research methods to generate a broader scientific understanding. However, Breitenbach and Choi argue that pluralism alone is not sufficient if scientific practitioners are following different methodologies and not working in communication and collaboration with one another. According to Breitenbach and Choi, '[a]dopting different approaches is beneficial only if the individuals are communicating and cooperating enough that it can be said that they are together trying to solve the problem using a plurality of approaches'.[87]

Unified pluralism does not defend any pre-established notion of the unity of science, but simply opposes the tendency of pluralist methodologies to deny the possibility that science could be unified. Breitenbach and Choi argue that these methodologies do not consider the importance of communication and cooperation, which results in an inability to recognise the possible ways that nature could be unified. In contrast, unified pluralism aims to preserve the ideal of the unity of science without presupposing its specific form prior to scientific evidence. In principle, this unity could comprise a single homogenous reductive account of science, or a systematic collection of heterogeneous laws, but 'each of these parts would have to, in virtue of its heterogeneity, play a unique, necessary and specifiable role in contributing to our unified understanding'.[88]

The motivation to preserve a methodology of science that does not reject the possibility of discovering a unified science is important because it recognises how the method of pluralism can make it impossible to discover this unity. In many cases, the piecemeal approach to science means that scientists rarely consider questions about how particular aspects of research might fit together beyond their field under a broader unified perspective. Yet Breitenbach's and Choi's account is problematic for reasons similar to those raised against Kitcher. Like Kitcher, they overlook the potential incompatibilities between Kant's regulative account of scientific theories and contemporary metaphysical accounts. Transcendental idealism can guide science by providing a framework that establishes the limits of knowledge this for scientific enquiry. Breitenbach and Choi place much of this responsibility on scientific practitioners and philosophers of science:

> We assume the results of different inquiries can be compared and contrasted with one another in order to add to a better understanding of the same natural world, understanding that, if it *could* be completed, would be an understanding of the natural world as a whole. Aiming to

unify our theoretical insights in this way is, we suggest, what it means to regard the pluralist unity of science as a regulative ideal.[89]

By focusing on the results of science, rather than the method, their account does not consider how scientific methods can become barriers to collaboration. For example, consider the philosophies of scientific method developed by Kuhn, Lakatos and Feyerabend discussed in Chapter 1. These accounts explained how scientific theories have various inbuilt mechanisms that are directed towards their own self-preservation.[90] While unified pluralism allows for the possibility of a unified account of science to develop, in practice, the methodologies of science may have inbuilt mechanisms that create potential difficulties for realising this unity. To ensure the development of science promotes cooperation and communication, it would be necessary to provide a framework agreed by all relevant parties defining the boundaries for scientific enquiry.

From a Kantian perspective, many disagreements within the philosophy of science arise from scientists adopting transcendentally realist principles. Kant emphasised that speculative disputes between transcendental realists were essentially irresolvable because they did not relate to any possible experience. According to Kant:

> how can two people conduct a dispute about a matter the reality of which neither of them can exhibit in an actual or even in a merely possible experience, about the idea of which he only broods in order to bring forth from it something **more** than an idea, namely the actuality of the object itself?[91]

For Kant, we must be sure that potential disagreements are based on disputes that can in principle be resolved; the strict regulation of the boundaries of knowledge is Kant's attempt to secure this. For instance, the incompatibility between Bhaskar's or Cartwright's philosophies of science related to different transcendental realist assumptions within their accounts.[92] Neither Bhaskar nor Cartwright opposed the idea of the unity of science in principle; however, they disagreed about how to understand this unity.

For Bhaskar, the possibility of science required that we presuppose the existence of intransitive transfactual laws hold uniform across reality, whereas scientific experimentation can only investigate these laws in their transitive context. Nature does not demonstrate law-like regularity at the empirical level, but this does not entail that nature in itself is not governed by intransitive laws. In contrast, Cartwright argued that there is no empirical justification for the belief in universal regular laws. She argued that laws should be understood as context-specific, relating to the experimental conditions or nomological machines that establish local regularity by assembling causal capacities in a particular way.

Transcendental idealism can offer a resolution to these accounts because it reveals how their incompatible appeals to the metaphysical structure of science presuppose that the conditions of objects in themselves can be derived from knowledge of appearances of objects. Both agree that laws do not express uniform law-like regularity at the level of appearance, yet they disagree about the implications this has for the metaphysical status of laws. Transcendental idealism preserves the unity of science as a regulative ideal without demanding that this unity is an ontological condition of objects independent of experience. The cost of this resolution is that philosophers of science must concede that they lack justification for deriving metaphysical claims regarding the status of laws from appearances.

Potochnik develops a more critical account, which potentially denies both the metaphysical status of scientific laws and the potential for these laws to achieve collaborative unity.[93] Potochnik's account of the role and importance of rampant idealisations in contemporary philosophy of science demonstrates how the models used by scientists are generally incompatible with each other. This becomes extremely problematic when we regard these models as relating to metaphysical accounts. She argues that scientists use idealisations to help them identify causal patterns that would otherwise remain undetectable. These idealisations should be understood as ways of furthering scientific understanding rather than discovering scientific truth. We should acknowledge the various human-centric factors that explicitly and implicitly shape scientific development. This presents a significant obstacle to our ability to generate any metaphysical knowledge by simply appealing to scientific findings. Moreover, it is an even greater obstacle for the idea that scientific findings will be compatible in the way outlined by unified pluralism because 'the various aims of science are in tension with one another, in the sense that they motivate different idealized products'.[94]

Potochnik's account highlights the previous concerns raised against the account of unified pluralism. In order to defend against the potential for incompatible idealisations arising, whether they are metaphysical or methodological, first requires scientists and philosophers of science to agree on the boundaries of scientific knowledge and practice. One way to establish agreement would be to endorse certain tenets of transcendental idealism. On the one hand, this would encourage science to develop in a cumulative and collaborative way, which avoids the possibility of scientists developing incompatible idealisations. On the other hand, Potochnik would most likely find this to be an unnecessary imposition on scientific practice. The reason that scientists can identify causal patterns in nature is because they deploy these idealisations. Essentially, science is not set up in a way that promotes the discovery of metaphysical truths or scientific unity. Potochnik claims that her account:

> undermines the idea that scientific products are converging – or will converge at some future point in time – and will then provide a unified account of the world. A diverse array of tools for piecemeal prediction, explanation, policy guidance, and so on holds little promise of direct metaphysical import.[95]

In summary, the support that Kant can offer contemporary philosophy of science comes at a cost. For Kitcher, it requires us to replace certain aspects of Kant's transcendental idealism that are incompatible with contemporary philosophy of science, such as his commitment to *a priority*. In contrast, Breitenbach and Choi argue that Kant's account of regulative unity would help direct scientific research towards communicative and collaborative ends. This would entail that scientific research must be consistent with the broader metaphysical limitations on knowledge imposed by transcendental idealism. Unified pluralism provides an outline of how a regulative principle of scientific unity could promote collaboration between scientists working in diverse fields. However, some philosophers of science such as Potochnik argue that the idea of scientific unity is inconsistent with contemporary scientific methodology.

3.2.2 Consilience and reductivism in philosophy of science

Consilience has been an important principle for philosophers of science who have used the principle of the unity of science as motivation for attempts to reduce the domain of ethics to biological explanations. I argue that these accounts appeal to a conception of consilience that significantly differs from Whewell's original formulation, which appealed to consilience as something essentially unexpected. In contrast, contemporary accounts of consilience tend to appeal to a metaphysical demand for the unity of science, which is largely detached from empirical justification. These more recent appeals to consilience offer a very different account from its original Whewellian formulation. I argue that these accounts transform consilience into a speculative assumption of the complete unity of nature, which goes beyond the remit of empirical justification. Although Whewell justified consilience through his theological resolution, he maintained that our knowledge was significantly limited in comparison's to God's knowledge.

Mackie's criticism against moral objectivity rests on fundamental assumptions about the underlying unity of science. He criticised the notion of objective values in his well-known argument from queerness. According to Mackie, any moral theory committed to an account of objective values requires that such values are simultaneously objective and prescriptive. He argues that if such entities were to exist, they would be 'utterly different from anything else in the universe' and we would require 'some special faculty of moral perception or intuition, utterly different from our ordinary ways of knowing'.[96] In this context, it is not the argument from queerness that is of primary interest, but Mackie's alternative biological explanation for moral behaviour.

Mackie aligns himself to empiricism as he argues that experience provides direct access to the physical world. Our 'ordinary way of knowing' entails that we have objective knowledge of the content of experience. Importantly, the argument from queerness is not directed against non-objective moral prescriptions: '[w]e all have moral feelings … which we therefore try to encourage and develop or to oppose.'[97] The reason for associating Mackie with the principle of consilience is his faith in future developments in biology to explain in terms of evolutionary mechanisms. According to Mackie:

[T]he ordinary evolutionary pressures, the differential survival of groups in which such sentiments are stronger, either as inherited psychological tendencies or as socially maintained traditions, will help to explain why such sentiments become strong and widespread.[98]

Mackie's argument against objective moral values is further supported by the underlying assumption that biological evolutionary pressures will eventually offer an alternative explanation for morality. The investigation of moral behaviour as originating from biological processes is most widely associated with Wilson's *Sociobiology*. Wilson describes his account as taking morality from the hands of philosophers and approaching it from the perspective of biology.[99] For him, the difference between the philosophical and biological approaches towards morality concerns the way that each regards its origin. Philosophy has approached the brain 'as though that organ must be treated as a black box'.[100] Wilson argues this has led philosophy to generally adopt a transcendental[101] perspective on ethics that regards morality as existing outside of nature, whereas the biologists adopt an empiricist perspective on morality as emerging from evolutionary pressures. According to Wilson, '[c]ausal explanations of brain activity and evolution, while imperfect, already cover the most facts known about moral behavior with the greatest accuracy and the smallest number of freestanding assumptions'.[102] For Wilson, the discovery of the biological foundations of ethics would reconcile the division between the humanities and sciences and contribute towards a universal consilience of nature.

The principle of consilience is central to Wilson's argument. However, his account differs significantly from Whewell's original formulation. Wilson's application of consilience supports an ontologically reductionist account of science. He emphasises that biology will offer an account of morality that is reducible to evolutionary principles. Wilson argues that if this biological explanation of ethics fails, then the possibility of a universal consilience also fails. According to Wilson, 'if evidence contradicts empiricism in any part, universal consilience fails and the division between science and humanities will remain permanent all the way to their foundations'.[103] In contrast, Whewell's appeal to consilience as an unexpected consequence is crucial for understanding its explanatory power. According to Whewell, '[w]hen principles in some instances have proved sufficient to give an unexpected explanation of facts, the delighted reader is content to accept as true all other deductions from the principle'.[104] Against Wilson, this does not entail that any single theory must explain every kind of fact. Nor does it entail that if consilience is not achieved in a particular case, then it could not be achieved in a different case. If a biological explanation of morality is impossible, then biological explanations might still achieve consilience when applied to other kinds of fact.

Wilson's expectation of a biological explanation of morality could not be greater; the possibility of consilience depends on a biological explanation of ethics. Wilson regards consilience as the method and metaphysics by which science discovers the single unified ontological foundations of the laws of nature. The belief in consilience is metaphysical insofar as it is grounded on the adverse consequences that would arise for science if reality did not manifest consilience at an ontological level. In contrast, Whewell understands consilience primarily as a methodological procedure. It is not the metaphysical presupposition of a reductive science, but rather a methodology for testing hypotheses. Larry Laudan explains that this has tended to be overlooked by Whewell scholarship at the cost of significantly misinterpreting his account. According to Laudan:

> Whewell's aim in stressing the consilience of inductions is not to maximize content, but to maximize the confirmation of an hypothesis. Of course, Whewell did believe that in the progressive growth of science, we advance towards theories of greater scope, range and generality. But (and this is crucial) increased generality is only a gain insofar as that greater generality is *experimentally confirmed*. Consilience is, effectively, a criterion of acceptability which stipulates that those hypotheses are most worthy of belief and acceptance which pass empirical tests.[105]

The increase of knowledge is a by-product of consilience. The primary effect of consilience is to demonstrate the correctness of a theory because it can be applied to a different case. If a biological explanation can be applied to facts belonging to a different kind, it increases the likelihood that the biological explanation is correct.

The difference between consilience in its reductive metaphysical and Whewellian forms is that the former expects to explain everything in accordance with a single metaphysical framework. Whewell's original formulation conceived of consilience as an unexpected result of scientific practice, which helped to consolidate the empirical evidence for a theory. According to Dupré, metaphysical assumptions about the unity of nature 'are not the most plausible explanations, and certainly not the responses that should appeal to a committed empiricism'.[106] These metaphysical assumptions lack empirical support because the belief that all scientific phenomena will be reducible to a unified explanation takes precedent over the current empirical evidence supporting or opposing that belief. Dupré's alternative suggestion is to recognise that nature is not unified; rather, it comprises ontological processes, which are more adequately explored through pluralist approaches to science. Of course, process ontology is also a metaphysical hypothesis. Dupré argues that the distinction between metaphysics and epistemology is not one of kind but rather one of degree. 'If it turns out that process is indeed the right concept to make sense of nature, then this is as good a reason as we can expect for taking nature to be ontologically composed of process'.[107] This dispute between ontological reductivism and processualism in this context is a debate that goes beyond empirical confirmation. From a Kantian perspective, both accounts propose irreconcilable transcendental realist perspectives.

Conclusion

Kant influenced the development of science in the British Isles in a variety of ways. His account of the active powers of the mind, knowledge as derived from two irreducible sources, and the unity of science, were influential for Whewell's philosophy of science. Yet Whewell deviated from transcendental idealism because of his dissatisfaction with the implications of transcendental idealism for scientific knowledge. Whewell argued that scientific knowledge related directly to objects in themselves, rather than to objects as they appear to us. Whewell appealed to the existence of God to justify these aspects of his philosophy. The role that the existence of God plays within Whewell's account of science is concerning given the continued significance of the unity of science within the philosophy of science. Whewell's appeal to God allowed him to transform the unity of science from a Kantian regulative demand of reason to a constitutive proof of a theory. Recent philosophers who appealed to consilience and the unity of science have not provided alternative reasons for continuing to search for greater unity within science.

Recent interpretations of Kant have explored the potential benefits of Kant's philosophy for contemporary philosophy of science. Generally, these accounts tend to minimise the significance of the distinction between constitutive and regulative principles for philosophy of science. Kitcher developed a Kant-inspired account of science, which argued that scientific theories are rational projections of order and that we must select from the best available theories. Similarly, Breitenbach and Choi argued that Kant's account of the regulative unity of science could offer support to contemporary pluralist methodologies. Kant's critical philosophy could only distinguish scientific theories that are compatible with transcendental idealism from those that are not. These accounts overlook how developing an account of science that is compatible with Kant's philosophy requires an examination of the methodological and metaphysical assumptions of science. For Kant, it is not possible for multiple theories that are consistent with transcendental idealism to offer conflicting accounts because reason cannot conflict with itself if it is deployed correctly. Many disagreements between contemporary philosophers of science relate to metaphysical disagreements that cannot be resolved by appealing to any experiment or experience. From the perspective of transcendental idealism, these disagreements arise because philosophers of science claim to know metaphysical truths that transcend the limit of knowledge. Irresolvable disagreements at the metaphysical level are barriers to the communicative and collaborative potential of science. In addition to agreeing on the focus of scientific research, scientific practitioners must also agree on the limits of scientific knowledge to avoid falling into irresolvable disputes. Transcendental idealism offers potential guidance to philosophers of science to develop their theories in such a way that encourages science to collaborate to identify and resolve issues in philosophy of science from a variety of perspectives.

4 • Whewell's Influence on Darwin and the Role of Design for the Organism

Introduction

The two significant sources of influence on the scientific method that Darwin developed in *On the Origin of Species* are from John Herschel and Whewell. In the first section I argue that their influences on Darwin relate to logically distinct arguments in Darwin's *Origin* drawing from Elliot Sober's analysis of *Origin*. Darwin argues that the vestigial traits of organisms provide insight into their biological ancestry. This bares a close resemblance to Whewell's account of consilience discussed in the previous chapter because it draws on parsimony as evidence for the correctness of a theory. Moreover, Darwin believed that his argument for common ancestry was especially strong because it drew from evidence across various biological disciplines. In contrast, Darwin's evidence in support of the argument for natural selection is more closely related to Herschel's account, which emphasises the importance of analogical reasoning. The evidence supporting each of these arguments is in opposition. Evidence for natural selection identifies traits that are adapted to increase fitness to certain environmental niches, whereas evidence for common ancestry identifies traits between different species that have no fitness benefit.

In the second section I examine the importance of design in Darwin's theory. Ruse has argued that Darwin would have agreed with Kant's account of the organism, or at least that certain features of organisms can only be understood in accordance with design. Ruse's appeal to Kant is problematic for two reasons. First, there is no evidence to suggest that Darwin read Kant's account of the organism in the *Critique of the Power of Judgment*. Second, some aspects of Kant's philosophy are incompatible with Darwin's account of natural selection. For Darwin, artificial selection served as an analogical basis for understanding natural selection. Darwin's analogy between artificial and natural selection bears a closer similarity to aspects of Paley's theological argument for divine design. In contrast, Kant argued that external purposiveness – which is similar to artificial selection – is fundamentally different from the internal purposiveness that we perceive in organisms. We must judge organisms as possessing a formative force, whereas artefacts possess only a motive force. Kant's account of the organism differs significantly from the accounts by Darwin and Paley because Kant opposes the idea that we can regard the design of organisms as analogous to the design of artefacts.

In the third section I investigate how design has continued to be a central principle for the development of biology. In 1979, Gould and Lewontin criticised adapationist thinking within biology. They argued that biologists tend to overemphasise the role of fitness at the cost of examining other factors that are relevant to biological development. The idea that biologists have placed too much emphasis on fitness exposes the role of judgement in understanding biological development. The role of judgement in biology has been considered from a Kantian perspective by Matthew Ratcliffe. He argues that Dennett's account of an intentional stance requires a non-naturalistic foundation to explain how it is possible to conceive of nature *as if* it were governed by intentions. Dennett presupposes a framework similar to Kant's account of teleological judgement, and it follows that Dennett's account requires a non-naturalistic standpoint. If we must consider organisms through the lens of design, these discussions often relate back to Kant's philosophy in significant ways. In contrast, some contemporary philosophers of biology such as Nicholson argue that it is not necessary to understand organisms in accordance with design at all. For Nicholson, organisms should be understood in the context of non-designed dissipative structures rather than machines or artefacts. I argue that the problem of design is resolved at the cost of distinguishing between living and non-living dissipative structures. Kant's account of organisms helps us to understand how the rejection of design results in the inability to establish the conditions that distinguish living entities from non-living entities.

4.1 Whewell's influence on Darwin's *On the Origin of Species*

For Whewell, our ability to identify design in nature is evidence in support of nature having been designed and created by God. In contrast, Kant argued that we are significantly more limited in the conclusions that we can draw from our experience of the purposiveness of nature. Kant recognised how the appearance of design in nature naturally led us to infer divine design, yet he argued that we could not infer God's design from our ability to conceive of nature teleologically. Their differing stances on theology marks a significant difference between the role of design for Kant and Whewell. The two main arguments in Darwin's *Origin* are the argument from common ancestry and the argument for natural selection. Each of these arguments relate to differing prevalent methodologies of science in Darwin's time. The argument for common ancestry appealed to Whewell's account of consilience as the fundamental principle of science. In contrast, the argument for natural selection drew from Herschel's argument from analogy. Darwin argued that natural selection was analogous to artificial selection.

4.1.1 The role of design in Whewell's philosophy

Whewell's account of biology drew from Kant's discussion of teleological judgement; however, he argued that this supported a theological conception of design in nature. In contrast, Kant argued that purposiveness can only relate to our judgement and can take two forms: either internal or external. Exploring these aspects of Kant's philosophy makes it possible to identify some fundamental differences between the accounts of the organism from Kant and Whewell.

Whewell's conception of biology was indebted to Kant's account of the organism. He argued that the fundamental idea required for biology was the Kantian definition of organisation in accordance with final causes; 'an organised product of nature is that in which all the parts are mutually ends and means' (Kant, cited in Whewell).[1] Life could not be reduced to any single function; rather, it must be understood as a system of functions. Our understanding of biological organisms includes the idea of final causes, as they cannot be separated from the inherent notions of end, purpose and design. Whewell argued that this cannot be deduced from the phenomena, 'but is *assumed* as the only condition under which we can reason on such subjects at all'.[2] In other words, our experience of final causes is not given externally as a fact but is generated from within the subject as an idea. This is an inherently Kantian aspect of Whewell's account as Kant also rejected any possibility that we could derive our idea of final cause from nature. According to Kant, 'it is absolutely impossible for us to draw from nature itself any explanatory grounds for purposive connections'.[3] For both Kant and Whewell, final causes are necessary for the possibility of comprehending biological phenomena. Moreover, they are not something that can be sufficiently explained from our experience of purposes in nature.

The difference between their accounts of final causes is evident from how they each justify our ability to perceive nature in accordance with final causes. For Kant, there are two ways that we can judge our experiences of nature. These are either in accordance with mechanical or teleological principles. Kant presents this as the antinomy of the power of judgement. According to Kant, the thesis of this antinomy is that '[a]ll generation of material things and their forms must be judged as possible in accordance with merely mechanical laws'.[4] The antithesis is that '[s]ome products of material nature cannot be judged as possible according to merely mechanical laws'.[5] This antinomy orients our investigation of nature by prescribing us to first investigate nature from a mechanical perspective. When nature cannot be sufficiently understood in accordance with mechanical laws, then we are permitted to judge nature in accordance with teleological laws. This antinomy is regulative because it refers to the way that we judge nature rather than making claims about the way that nature is independent of judgement. The regulative – rather than constitutive – status of this antinomy allows us to regard these two ways of orientating ourselves towards nature as complementary to one another, rather than contradictory. According to Philippe Huneman:

> mechanism and teleology can be conceived of as two complementary explanatory stances. The first one uncovers processes at work in all of nature, and therefore it is not proper to biology. However, when facing a particular process taking place in an organism, it does not answer questions such as: why is this process here? ... This second explanatory stance clearly does not concern *non organized* entities.[6]

Huneman correctly identifies the importance of the complementarity between mechanical and teleological judgements as regulative principles. However, there are inherent problems with Huneman's relegation of teleological judgements to a type of question, a 'why' rather than a 'how' question, that can be asked of nature. Huneman does not consider the significance of Kant's claim that teleological judgement is necessary for the possibility of any experience of organisms. In other words, it is not merely that teleological judgement allows us to direct certain questions towards experience, which we would otherwise remain unaware of; rather, the very conception of organisms requires that we perceive aspects of our experience of the world *as if* they are directed towards final causes.

Whewell's account of final causes is significantly similar to Kant's, as both regarded final causes as necessary for us to conceive organisms, but Whewell was not satisfied with the argument that these features of organisms could be understood as relating merely to judgement. Instead of prioritising mechanical explanations over teleological explanations of nature, Whewell argued that metaphorical language was not limited to teleological explanations as mechanical scientific explanations also appealed to metaphorical language. Terms often used in mechanical sciences, such as energy and effort, likewise imply volition and animated action. According to Whewell:

> We endeavour in vain to conduct our mechanical reasonings without the aid of this idea, and must express it as we can. Just as little can we reason concerning organized beings without assuming that each part has its function, each function has its purpose.[7]

Kant limits the role of final causes to a last resort in scientific explanation for cases where mechanical explanations are not possible. Whewell justifies final causes, and other metaphors in science that imply the existence of final causes, by arguing they are necessary for scientific understanding. The role of metaphor within science has continued to create tensions within science. Metaphors are indispensable to our ability to do science – to this effect, Lewontin argued that '[i]t is not possible to do the work of science without using language that is filled with metaphors'.[8] He warned that the price of using metaphors in science is eternal vigilance against the possibility of mistaking the metaphors for the real objects of scientific interest.

Whewell is not appropriately vigilant against the implications that arise from the metaphor of design. Whewell considers the indispensability of the metaphor of design as evidence for God's existence. According to Whewell, '[i]f men really can discern, and cannot help but discerning, a design in certain portions of the works of creation, this perception is the soundest and most satisfactory ground for the conviction to which it leads'.[9] The conviction is that God exists as the creator and designer of nature. Moreover, this design is not limited to biological organised beings. In fact, he argues that even mechanistic scientific laws can be understood in teleological terms. Whewell explains that the Copernican revolution did not cast doubt on the idea that the sun is intended to offer warmth and vitality to plants and animals because of the discovery that the Earth revolves around the sun. According to Whewell:

> Final causes, if they appear driven further from us by such an extension of our views, embrace us only with a vaster and more majestic circuit: instead of a few threads connecting some detached objects, they become a stupendous net-work, which is wound round and round the universal frame of things.[10]

Whewell's argument that design in nature is evidence for divine creation reveals the potentially expansive way that nature can be understood in accordance with design. Whewell acknowledged that the way design is comprehended in the case of the solar system is different from the way design is comprehended for organic entities; in the former, it is comprehended at the level of laws rather than at the level of individual entities. According to Whewell, '[t]he principle of design changes its mode of application indeed, but it loses none of its force'.[11]

Kant's account of teleological judgement is more critical than Whewell's when it comes to locating the source of design outside of our own capacities. Kant recognised that many aspects of our experience of nature seem to exemplify design, but we are not justified to claim that the origin of this design is either entirely natural or supernatural. He distinguishes between relative (or external) purposes and natural (or internal) purposes. Relative purposes refer to aspects of nature where design relates to usefulness or advantageousness for humans or other creatures. Kant explains that in these cases nature exemplifies purpose and design for non-human animals in a merely relative sense. According to Kant, 'if cattle, sheep, horses, etc. were even to exist in the world, then there had to be grass on the earth'.[12] The ways that humans use nature for their own ends are also examples of relative purposes. Kant explains how humans use nature for 'foolish ends' such as decorating and dying their clothing with feathers, soils and juices. Humans also use nature for rational ends such as riding horses and using swine and oxen to plough fields. In these cases, the purpose in question is always external and directed towards another aspect of nature. Relative purposes cannot become natural purposes because we cannot know that grass necessarily exists to feed cattle, or that horses necessarily exist to fulfil our purpose of horse riding. Relative purposes can only become natural purposes if we know the entity that uses another for its own ends must necessarily exists. According to Kant:

> Only if one assumes that human beings have to live on the earth would there also have to be at least no lack of the means without which they could not subsist ... those things in nature which are indispensable for this purpose would also have to be regarded as natural ends.[13]

Knowledge of our own necessary existence is impossible. All we know is that we do exist, but we do not know that we must exist. Kant concludes that relative purposes only hypothetically indicate the existence of natural purposes but cannot provide sufficient justification for them. Relative purposes cannot offer any justification for why an entity must exist, which is the essential condition of natural purposes. A relative purpose only relates to how humans and other animals can utilise other aspects of nature to fulfil their needs and wants, which does not provide any insight into the purpose as causally responsible for the existence of that entity. Therefore, it is not possible to conclude our use of nature and other animals corresponds to their absolute purposiveness.

In contrast to Kant, Whewell aimed to show that humankind was predestined to exist in nature, and the ability for us to spy design in nature was evidence of this divine designer. According to Richard Yeo, Whewell 'believed that man's ability to understand the laws (and thoughts) of God implied a threefold harmony between the mind of God, the mind of man and the laws of nature'.[14] In this context, Whewell's theological commitment allows him to conceive of the appearance of design and purpose in nature as part of God's plan.

The difference between the accounts of biology developed by Whewell and Kant is important for understanding their significance for Darwin. Darwin's account of natural selection developed an alternative explanation for biological development that was not significantly influenced by the accounts of biology developed by Kant or Whewell. Whewell's conception of consilience discussed in the previous chapter was a much greater influence on Darwin's theory. In this sense, Kant's indirect influence on Darwin was his influence on the development of Whewell's philosophy of science, specifically his conception of consilience. It is indirect because Whewell transformed various aspects of Kant's philosophy to create a methodology that aligned with the account of science at Whewell's time. Whewell argued that the demonstration of consilience increased our knowledge of truths about the world independent of experience. He appealed to the history of science as evidence of consilience, as he argued that once consilience had been demonstrated it had never been proven false. In contrast, Kant's transcendental justification for unity was a regulative demand of reason. While Kant influenced the development of Whewell's philosophy, Whewell transformed aspects of Kant's account so that he could establish a philosophy of science that related to things in themselves.[15]

4.1.2 The influence of Whewell and Herschel on the arguments of Darwin's Origin

Darwin famously concluded the *Origin*[16] with the claim that it contained 'one long argument'.[17] Various philosophers of biology have questioned this claim. For instance, Mayr argues that Darwin's theory contained five sub-theories,[18] and possibly more depending on divisions that individuals may prefer to impose. In contrast, Sober has argued that the *Origin* contains two logically distinct arguments: the argument for natural selection and the argument from common ancestry. There are important differences between the argumentative strategies that underpin each of the arguments identified by Sober, and these related to broader differences between the methodologies proposed by Herschel and Whewell.

According to the argument from common ancestry, certain traits do not increase the overall fitness of an organism and therefore cannot be regarded as a direct consequence of natural selection. These vestigial traits, which often have no current use, can be used as evidence for common ancestors because there is a greater probability that any species possessing this trait share an evolutionary lineage with other species that also possess it. Sober considers the evidence for a common ancestor between humans and monkeys on the basis that both have tail bones:

> It is false that human beings and monkeys must both have tailbones if they have a common ancestor. It is also false that they cannot both have tail bones if they lack a common ancestor. What is true is that the probability of this similarity is greater under the common ancestry hypothesis.[19]

Evidence for common ancestry does not amount to certainty for each case. Rather, it is merely the most probable explanation for this similarity. The hypothesis is most probable in virtue that tailbones do not have a current use in human beings. Darwin appeals to many different areas of biology as evidence for examples of common ancestors. For instance, Darwin considers cases where there are similarities between different biological groups during embryological development that then disappear once members of the groups reach maturation. Darwin argues that such similarities are evidence for common ancestry:

> In two groups of animals, however much they may at present differ from each other in structure and habits, if they pass through the same or similar embryonic stages, we may feel assured that they have both descended from the same or nearly similar parents ... Thus, community in embryonic structure reveals community of descent.[20]

Common ancestry provides clearer evidence for Darwin's argument that emerging species descend from other species. If organisms from vastly different species both possess a trait that no longer increases the fitness for some of these trait-bearers, then this offers support to the probability that these species are related to one another. In contrast, if a trait is present in multiple species and it contributes to an increase of fitness for all trait-bearers, then it is likely that this trait does not reveal a common ancestor. It is more probable that such traits emerged because of selective forces that cause biological entities to 'become adapted to similar conditions, and thus assume a close external resemblance; but such resemblances will not reveal – will rather tend to conceal their blood – relationship to their proper lines of descent'.[21]

Sober raises the question of whether Darwin wrote the *Origin* backwards because the greater evidential support for Darwin's account speciation comes from examples of common ancestry rather than natural selection. Sober concludes that 'the book is in the right causal order, but evidentially, the book is backwards'.[22] The two arguments stand in tension with one another. The argument for common ancestry identifies traits without a current a function and uses this as evidence for the biological lineage between species. Whereas the argument for natural selection demonstrates how species adapt traits that increase fitness for trait bearers. One reason that the argument for natural selection is evidentially weaker than the argument for common ancestry is because support for the theory is established by analogy with artificial selection. Artificial selection in this context refers to our ability to change the traits of species through selecting those traits that were beneficial or desirable for us.

Darwin's use of analogical arguments and consilience as argumentative strategies within the *Origin* relates to differing methodological accounts of science at his time. The two prevalent methodologies of science during the development of Darwin's *Origin* were developed by Herschel and Whewell. The predominant interpretation is that Darwin's use of the principle of consilience is evidence of Whewell's influence, and the Herschel's influence was less significant. Ruse has strongly advocated this interpretation, which has been influential on Whewell scholarship.[23] Darwin's appeal to various different disciplines in biology as supporting common ancestry demonstrates that Darwin placed an emphasis on the role of consilience within his account. However, it is misleading to say that Whewell rather than Herschel influenced Darwin, as his account draws from the methodologies of both in different ways.

Ruse's explanation of the influences of Hershel and Whewell on Darwin has changed over time. Examining these changes reveals some important differences between the methodologies of Herschel and Whewell. In an early essay on the subject, Ruse presented their views in combination under the more general 'Herschel–Whewell philosophy of science'.[24] In response, Paul Thagard explained how the procedures of establishing *vera causa* principles (or true causes) were significantly different for Herschel and Whewell. For Herschel, a *vera causa* principle is derived by identifying analogies that reveal an underlying natural cause, whereas Whewell argued that true causes are revealed when a theory achieved a consilience of inductions.

This distinction is important because the influence of each account of *vera causa* can be identified within different aspects of Darwin's theory. Darwin's discussion of the analogy between artificial and natural selection bares a close resemblance to Herschel's emphasis on analogy. According to Herschel, '[i]f the analogy of two phenomena be very close and striking, while, at the same time, the cause of one is very obvious, it becomes scarcely possible to refuse to admit an action of an analogous cause in the other'.[25] For example, Herschel suggests that swinging a stone in a sling in a circular motion is analogous to the moon orbiting the Earth. Gravity is the analogous cause to the sling. Natural selection is an analogous cause to the ability for humans to artificially select traits.

Thagard argues that Darwin adopted a Whewellian account of *vera causa* as consilience: 'all that mattered is that it explained several classes of facts and thus achieved a consilience of inductions. Hence Darwin did not need the analogy between natural and artificial selection to show that natural selection is a true cause'.[26] This exchange between Ruse and Thagard led to important clarifications relating to the differences between Whewell and Hershel. Ruse responded to Thagard by conceding the difference between their accounts of *vera causa*. He explains this difference as follows:

> Herschel, whilst recognizing the importance of a consilience, did not in itself elevate it to the status of something showing a *vera causa*; rather, he kept always with his analogies for the defining mark of a *vera causa*. Whewell on the other hand argued that a Herschellian *vera causa* is no *vera causa*, and that the only real *vera causa* is a cause at the centre of a consilience.[27]

Against Thagard, Ruse argues that it is misleading to simply argue that either Whewell or Herschel influenced Darwin because they offer incompatible accounts of establishing *vera causa* principles. It is not necessary to set the influences of Herschel and Whewell against one another as Darwin's appeal to consilience is compatible with both. This potentially underplays the difference between Herschel and Whewell. For instance, Whewell's and Herschel's accounts of *vera causa* are, at times, contradictory.[28]

An argument against Whewell's influence on Darwin has come from the historical analysis of Darwin's account. Whewell could not have been significantly influential on the development of the structure of Darwin's argument, as it did not alter from 1838, which was prior to the publication of Whewell's account of consilience.[29] Yet, there are several passages of Darwin that are only compatible with the Whewellian account of *vera causa* as requiring a consilience of inductions. For instance, Darwin concludes that if several classes of fact can be explained in accordance with common ancestry and transmutation, then he 'should without hesitation adopt this view, even if it were unsupported by other facts or arguments'.[30] This passage is evidence for the significance of Whewell's account of consilience for the argumentative strategies in the *Origin*, even if it only corroborated existing germinal ideas that Darwin had developed prior to knowing about Whewell's theory of consilience.

The reception of Darwin's theory also indicates that the Whewellian aspects of Darwin's argument made a greater impact on the reception of Darwin's account. The arguments for the transmutation of the species and common ancestry received a greater level of acceptance than the argument for natural selection. Kenneth Waters argues that this would be inconsistent if those sympathetic to Darwin's theory approached Darwin from a Herschelian methodology. According to Waters, '[i]f natural selection is removed, the alleged *vera causa* vanishes, the Herschelian argument collapses, and there is no reason for accepting transmutation or common descent'.[31] Many were cautious of the analogy between natural and artificial selection, but this analogy was central to the Herschelian elements of Darwin's account. Hence, regardless of the dispute towards Darwin's own intentions for the methodology of the *Origin*, there is evidence to suggest that its reception favoured a Whewellian, rather than a Herschelian, interpretation.

The importance of Whewell's account as an argumentative strategy for the *Origin* raises many important questions given the complex relationship between Whewell and Kant discussed in the previous chapter. For Kant, the unity of nature is a regulative demand of reason that is necessary for the possibility of scientific enquiry. Whewell deviated from Kant's philosophy because he rejected both Kant's restriction of knowledge to the fixed forms of the categories and the rejection of knowledge of things in themselves. In contrast, Whewell claims that knowledge is the product of the fundamental antithesis between thoughts and things in themselves. Whewell's account ultimately depended on a theological ground for the resolution of the fundamental antithesis and his conception of consilience.

Regardless of whether Darwin derived his account of consilience directly from Whewell, Darwin's theory requires an explanation for why consilience is an explanatory virtue for a theory. Darwin provides no justification for why a theory that can be applied to a greater variety of facts should be preferable to a theory that has a smaller application. For Kant, this is justified in accordance with the regulative demand of reason,[32] whereas for Whewell, justification comes from an appeal to 'God's grand design'. Darwin's appeal to consilience is indebted to the theologically grounded conception of science developed by Whewell.

4.2 Darwin and design: the relation between artificial and natural selection in Darwin's *Origin*

The analogy between artificial and natural selection has remained one of the most controversial aspects of Darwin's account. Darwin drew from Herschel's conception of the *vera causa* to justify this aspect of his account, but the significance of the argument from analogy has led to disagreements within literature. Richards has argued that Darwin could not have intended this relation to be analogical because of the insurmountable differences between artificial and natural selection. In contrast, Ruse has argued that we should consider Darwin's account of natural selection in combination with Malthus's influence on Darwin. This helps us to appreciate the analogous relationship between artificial and natural selection despite the various ways that natural selection extends beyond artificial selection.

The theologian, William Paley, was a key figure in the development of Darwin's account of natural selection. Paley argued that organisms are evidence for divine design in an analogous way to assuming that a watch found on a heath is the product of intentional design. While Darwin did not explicitly argue that organisms are the product of divine design, his account accepted the analogy between organisms and artefacts that was central to Paley's argument. There is no evidence to suggest that Darwin was aware of Kant's account of the organism, but Kant does present important arguments against the analogy between organisms and artefacts. Considering Paley's influence on Darwin exposes how the theory of natural selection inherited a central assumption about natural design from Paley's theological analogy. Kant's criticisms against the relationship between the analogy between artefacts and organisms raises difficulties for recent interpretations of Darwin that appeal to Kant's conception of teleological judgement as offering support to Darwin's account of design.

4.2.1 The relation between artificial and natural selection

Darwin's discussion of artificial selection focuses on our power to identify traits and characteristics of animals and plants and breed only from those that best display these traits and characteristics to increase their frequency in future generations. According to Darwin, 'man can act only on external and visible characters: nature cares nothing for the appearance, except in so far as they may be useful to any being ... Man selects only for his own good; nature only for that of the being which she tends'.[33] The analogy between artificial and natural selection has been a topic of dispute within scholarship.

Many of Darwin's contemporaries regarded artificial selection as evidence against the causal efficacy of natural selection because it could not produce new species.[34] Natural selection entails an increase in fitness for organisms possessing traits that are selected, whereas artificial selection is not related to increasing fitness. For artificial selection, traits are selected in accordance with the breeder's intentions. According to Richards, 'in terms of fitness, the difference is one of kind. Natural selection favours fitness, while artificial selection opposes it'.[35] For this reason, Richards argues that Darwin's appeal to artificial selection is merely a psychological heuristic for illustrating some similarities with natural selection. There are too many incompatibilities between artificial and natural selection for it to be considered an argument from analogy: '[i]f we deny this, then it is unclear what would ever count against an analogical argument.'[36]

In contrast, Ruse has argued that there is an analogical argument between artificial and natural selection, but we must consider Thomas Malthus's influence on Darwin to understand this analogy. According to Ruse, '[b]efore reading Malthus, Darwin, if anything, stressed reasons why one who believed in evolution should not draw an analogy with domestic organisms'.[37] The pre-Malthusian Darwin accepted that the incompatibilities between the two forms of selection were so great that no analogy could be drawn between them. Ruse denies that this meant Darwin's theory developed independently of his awareness of artificial selection; rather, he had not yet found the principle that accounted for the differences between them. This principle was the struggle for existence that Darwin attributed to Malthus's doctrine when applied to all animal and vegetable kingdoms. Malthus revealed to Darwin that since more organisms are 'produced than can possibly survive, there must in every case be a struggle for existence'.[38] Reproduction increases as a geometrical ratio, but organisms can only be sustained at an arithmetical ratio, which is inevitably exceeded. On this principle, individuals that manifest variations that increase fitness are naturally selected. According to Darwin:

> as all organic beings are striving, it may be said, to seize on each place in the economy of nature, if any one species does not become modified and improved in a corresponding degree with its competitors; it will soon be exterminated.[39]

Only by considering artificial selection in combination with Malthus could Darwin account for both the similarities and the differences between artificial and natural selection. For artificial selection, traits are selected – either consciously or unconsciously – based on the use or value that they have for breeders. For natural selection, traits are selected that increase the fitness of individuals over their competitors. Artificial selection cannot offer an explanation for speciation as any changes are limited, whereas changes produced by the latter are unlimited and therefore speciation can be explained.[40] Natural selection exceeds the boundaries of artificial selection when considered in combination with a Malthusian perspective. According to Ruse, natural selection 'had not been thought to be so useful because it was, as it were, "dragged down" by the analogy from domestic organisms ... after Malthus, it was seen that natural selection can far outstrip its analogical relative'.[41]

Darwin avoids defining the essential features of the organism in part because he contextualises his discussion of organisms in the analogical relationship between artificial and natural selection. He regarded our knowledge of organisms as inherently unproblematic, possibly because they were already utilised for our needs with great success in cases of artificial selection. Darwin commented on pamphlets from many breeders in his notebooks. These pamphlets documented breeder's observations of the role of artificial selection within domesticated animals and livestock.[42] Darwin also avoided providing a specific definition of species; he merely asserts that 'every naturalist knows vaguely what he means when he speaks of a species. Generally, the term includes the unknown element of a distinct act of creation'.[43] Darwin's pragmatic approach to defining organisms and species draws from discussions between breeders who would not have been concerned with questions about the essential features of organisms. The ability to demarcate living from non-living entities, and the possibility to account for the diversity of organisms under a single definition, remains a live issue for contemporary philosophers of biology.[44]

The circumstances that led Darwin to investigate the origin of species and natural selection marks a significant difference between his and Kant's accounts. Kant's treatment of teleology in its external and internal forms differs from Darwin's account of artificial and natural selection as Kant was not concerned with our ability to select traits that induce physiological changes in progeny. Kant was not informed by the observations of breeders, nor was he concerned with the changes that could be produced by artificial or natural selection. His concern focused on how to understand the purposiveness required for the experience of organisms. As discussed in the first section of this chapter, Kant was also aware that humans often use organisms for their own ends, yet he did not consider how we could influence the physiological traits of animals based on selective breeding. Instead, his concern was whether we could generalise from the relative purposes that we can identify within nature to some idea of an absolute purpose within nature as a whole.

Darwin's analogy between artificial and natural selection does not consider a key philosophical problem raised by Kant; namely that, if the source of the intended use is external to the entity that is changed, then this cannot be an internal purpose. Darwin avoids this problem because he appeals to the phylogenetic changes that can be produced by both artificial and natural selection as evidence for their belonging to the same power. According to Darwin, 'if a feeble man can do much by his powers of artificial selection, I can see no limit to the amount of change ... which may be effected in the long course of time by nature's power of selection'.[45] Darwin does not regard the forces of artificial and natural selection as different in kind.

4.2.2 The distinction between organisms and artefacts for Darwin, Paley and Kant

There is a strong theological influence on the way that Darwin frames his account of natural selection. It is well known that Darwin had a strong admiration for Paley as a student at Cambridge. Like Kant, Paley recognised that the apparent design of organisms could not be the product of chance. According to Kant, 'nature, considered as a mere mechanism, could have formed itself in a thousand different ways without hitting precisely upon the unity in accordance with such a rule'.[46] Unlike Kant, Paley proposed that the only possible explanation for design in nature was theological. Paley's *Natural Theology* commences with the following thought experiment. Suppose that we find a watch on a heath. We would immediately regard the watch as the product of design as its parts are put together to serve a purpose. This would not be the case if we found a stone rather than a watch. Even if we have never seen a watch before, or the watch is in some way defective, we would still regard it as the product of a designer.[47] Paley then asks us to imagine that this watch has the unexpected property of producing another watch, like itself. According to Paley:

> Though it be now no longer probable, that the individual watch which our observer had found, was made immediately by the hand of an artificer, yet doth not this alteration in anywise affect the inference, that an artificer had been originally employed and concerned in the production.[48]

Paley thought experiment served as the basis for his argument that
'[t]here cannot be design without a designer',[49] which enabled him to
argue that the design behind biological organisms exhibited in nature
is the product of intelligent design. This watch possessing the capacity to produce other watches is analogous to the traits possessed by
organisms. Organisms are analogous to machines. Darwin's deviation from Paley is not the rejection of viewing nature as designed, but
rather his rejection of the claim that such design required a designer.
According to Francisco Ayala, 'Darwin's argument addresses the
same issues as Paley's: how to account for the adaptive configuration
of organisms, the obvious "design" of the parts to fulfil their particular functions'.[50] Hence, Darwin did not deny design in nature.

There is a strong similarity between Darwin's account of artificial selection and Paley's self-replicating watch. The watch, much
like the traits that are artificially selected by breeders, could never
arise by means of natural selection. In both cases, their functions
are not selected to increase their fitness to their environments. The
agency behind this kind of selection is external to the entity mani-
festing the mark of design.

In addition to Kant's criticism against the analogy between
artificial and natural selection, Kant would have also been critical towards Paley's thought experiment. Kant denied that finding
a shape drawn in the sand on a deserted beach is analogous to
our comprehension of organisms. Much like the watch, the shape
drawn in the sand can only be understood as a product of reason rather than nature.[51] The possibility that nature, unguided by
intentions, could have produced such a shape is infinitely small.
There is little doubt that the shape in the sand is the product of
an end, but it is not a natural end. Kant argues that a natural end
requires us to judge an entity itself as both the source and referent
of its own organisation.

Kant fissures the relationship between relative (or external)
ends and natural (or internal) ends. Artefacts only manifest purposes that are dependent on the plans of another, whereas the
purposiveness of organisms is judged as emerging from nature. We
create artefacts to fulfil certain purposes, which are the reason for
their existence. Artefacts do not possess the ability to regenerate
their parts or their whole. Hence, Kant denies the possibility that
a watch could possess the powers that are specific to organised
beings. According to Kant:

one wheel in the watch does not produce the other, and even less does one watch produce another, using for that purpose other matter (organizing it); hence it also cannot by itself replace parts that have been taken from it, or make good defects in its original construction by the addition of other parts, or somehow repair itself when it has fallen into disorder: all of which, by contrast, we can expect from organized nature. – An organized being is thus not a mere machine, for that has only a **motive** power, while the organized being possesses in itself a **formative** power.[52]

Kant is in agreement with Paley and Darwin insofar as the appearance of organisms exhibit design. However, he denies that any knowledge of the natural purposiveness of nature can be derived from our knowledge of external purposiveness or purposes exhibited by artefacts. Both Darwin and Paley sought to establish a common ground between artificial and natural purposes, albeit in different ways. Paley appealed to divine design in combination with the similarity between our own organisation and the organisation of machines, whereas Darwin argued that we understand the idea of an unlimited naturally selecting force if artificial selection is considered in combination with Malthus. In contrast, Kant sought to isolate these forms of purposiveness from one another. Kant differentiated our awareness of organisms from our awareness of other kinds of objects. He denies the coherence of Paley's example of the self-replicating watch because machines cannot reproduce their whole or their parts. For Kant, there are three ways that our judgements of objects relate to purpose; these are 'man-made artefacts, inorganic objects, and living organisms'.[53] Inorganic objects lack purpose because our experience of them does not require us to view their organisation in relation to a function they ought to achieve. According to Hannah Ginsborg, 'there is nothing which a stone ought to be. We may judge a stone to be sound or defective with respect to some particular human purpose ... [b]ut we cannot describe it ... as sound or defective *tout court*'.[54] In contrast, organisms and man-made artefacts are both judged in accordance with purpose as their functions help us to understand how they ought to be, and thus they can be considered as defective if they are malfunctioning.

There has been a tendency to overlook the essential dissimilarity between organisms and machines within Kant scholarship. For instance, Ginsborg has noted that Peter McLaughlin correctly identifies the difference in kind between these entities, but then underplays this difference in his further treatment of the antinomy between teleology and mechanism. According to McLaughlin's interpretation of Kant, we need to appeal to teleological explanations when our current mechanistic explanations are insufficient. However, this does not entail that it will not be possible to provide a mechanistic explanation at some point in the future, and thus eliminate the need to appeal to teleological judgements. According to McLaughlin:

> [Teleological judgement] never impedes the possibility of a later mechanistic explanation ... Whether or not, in a thing that we have to conceive as a natural purpose, an unimaginable, non-mechanical, real causality is active, we can never know with certainty.[55]

In this case, teleological judgements are required for judging aspects of nature that are currently inconceivable in accordance with mechanical judgement. It is important to distinguish two possible ways to understand mechanism in this context. Mechanism could refer either to a machine or to a law. This difference is significant because the machines do not exclude the need to appeal to teleology. According to Ginsborg, 'for Kant there is no less of a need for teleology in understanding a machine such as a watch, than there is in understanding an organism'.[56] A machine is teleological in the sense that it requires a purpose that is external to the machine; Kant refers to this as a motive power. Whereas organisms possess a formative power that is internal and responsible for self-organisation, repair and reproduction of the organism.

The requirement that teleology should explain organisms in accordance with purposes signifies a similarity between organisms and artefacts. However, it is possible that future developments of our knowledge will offer a mechanical explanation in place of the previous teleological explanation. The discovery of these mechanisms would entail that there is no longer any need to appeal to a partial similarity between organisms and machines as there would no longer be the need to appeal to teleology to explain organic entities. According to McLaughlin:

> Should it turn out that a phenomenon judged teleologically can be explained mechanistically, e.g. on the basis of newly discovered

empirical laws, no contradiction can arise between the new mechanistic explanation and the superseded teleological explanation; what was teleological in the old explanation becomes superfluous and what was mechanistic in the old explanation remains valid.[57]

One problem with McLaughlin's account is that if the possibility of judging organic entities requires that they are viewed as possessing a formative non-mechanistic drive, then, it is difficult to comprehend how it is possible to discover a new mechanistic law that would supersede the teleological explanation. The possibility of experiencing organisms requires that we view them as if they are governed by a formative teleological force, without this force it is not clear in what sense our experience would be an experience of an organism. McLaughlin concedes the possibility 'that a purely mechanical explanation of the organism may perhaps never be successful without abandoning mechanism as the ideal of explanation'.[58] This inability to explain organisms in terms of mechanical explanations is not a shortcoming of Kant's philosophy. Instead, it exposes an assumption at the foundation of biology regarding the similarity between organisms and artefacts or machines. According to Arthur Lovejoy, Kant would have opposed Darwin's purely mechanistic explanation as he 'was most of all hostile to the supposition that any of the phenomena of organic life can be completely explained mechanistically'.[59]

The association of Darwin with a mechanistic explanation of life is further complicated by some interpretations that suggest Darwin was not offering a mechanistic account. This has been central to the ongoing dispute between the interpretations of Darwin proposed by Ruse and Richards. For Ruse, Darwin was a mechanist. The influence of the industrial revolution on Darwin's thought led Darwin to consider natural selection through an industrial lens. According to Ruse, '[t]his means competition, it means Progress, and above all it means machines'.[60] Darwin regarded God as the 'Supreme Industrialist' who created the machine of natural selection. Once the machine has been created, God becomes a 'retired engineer'. Thus, Ruse argues that Darwin did not regard God as necessary for natural selection at every moment; 'God is not needed. At most we have metaphors'.[61]

Ruse acknowledges that Darwin never refers to Kant on the apparent or metaphorical design of nature, but he asserts that

Darwin would have agreed with Kant insofar as 'you cannot do biology without the metaphor. The point is that there is no implication of an Aristotelian vital force objectively out there in nature making for final causes. Final causes are our way of thinking about a mechanistic system'.[62] Ruse argues that we apply the metaphor of design to organisms to stimulate new avenues of scientific enquiry by considering the possible adaptive functions of traits. Biology would be very different if it did not utilise the metaphor of design and would exist only in a limited sense in fields such as embryology, physiology and classification:

> If we want a biology that is not interested in the reason why the stegosaurus has such a funny display on its back, is not intrigued by the peculiar shape of the trilobite lens, does not care why some butterflies mimic other butterflies, is unconcerned about the spirals of the sunflower, then presumably something can be done.[63]

For Ruse, metaphorical thinking allows us to consider biological organisms *as if* they are the product of design. Final causes are not a necessary condition for the comprehension of organisms; rather, they only make it possible to ask why biological entities seem to have certain functions. Once these final causes can be explained in terms of their underlying mechanisms, then these final causes can be eliminated from biology.

Ruse's appeal to Kant is inconsistent with Kant's own explanation of the organism. According to Richards, 'Ruse fails to take the role of metaphor … in science seriously. He assumes that they can be eliminated while leaving theory intact'.[64] For Kant, final causes are not merely heuristic in the sense that we can choose to apply them to our experiences of organisms to prompt a new way about why certain biological entities have developed particular functions.[65] Final causes are a precondition of experiencing an entity as an organism. According to Kant, the 'maxim of the reflecting power of judgment is essential for those products of nature which must be judged only as intentionally formed and thus not otherwise, in order to obtain even an experiential cognition of their internal constitution'.[66] In this sense, Ruse's appeal to Kant is misleading insofar as Kant's account of teleological judgement requires that final causes are necessary for the 'experiential cognition of the internal constitution' of biological entities. Without final causes, it would not be possible to distinguish organisms from other kinds of objects within experience.

Moreover, it is problematic to take Kant's account of teleological judgement in isolation from the context of his critical philosophy. Kant developed his account of teleological judgement following his concerns regarding the incompatibility between theoretical and practical philosophy. In this sense, Kant regarded teleological judgement as potentially providing a unifying bridge between his theoretical and practical philosophy.[67] Kant's account of freedom was central to his conception of teleological judgement. According to Guyer, 'it is only our awareness of the freedom of our own purposiveness that leads us to conceive of the purposiveness of organisms as necessitating a fundamental split between the teleological and mechanical views of nature'.[68]

Kant's conception of teleological judgement is clearly not suitable to help us understand Darwin's account of design. Ruse appeals to Kant to show how Darwin could preserve the role of teleology as a merely heuristic principle that could guide scientific enquiry. Yet, this is inconsistent with the original formulation of Kant's discussion of teleological judgement because his conception of organisms is essentially non-mechanical. While Kant argues that we must approach nature mechanically wherever possible, this does not entail that it will be possible to explain even a mere blade of grass without viewing it as the product of intentions.[69] In this sense, Kant's account of teleological judgement is not helpful for understanding Darwin.[70]

In contrast, Richards argues that Darwin was greatly influenced by German Romanticism, which revealed to Darwin that natural selection was essentially a creative process, rather than mechanistic. According to Richards, '[n]ature hardly operates like a clattering and wheezing Manchester mechanical loom, rather like a subtle and refined mind that can direct development in an altruistic and progressive way'.[71] Richards interprets Darwin as arguing that the ultimate goal of natural selection is human beings. The inherent design and purposiveness of natural selection demonstrated to Darwin that the biological laws are 'produced by an intelligent mind governing the universe'.

It is interesting that Richards's interpretation sets Darwin closer to Kant in some important respects. Kant argued that our capacity for teleological judgement required us to judge the final end of nature as 'nothing other than the human being under moral laws'.[72] Teleological judgement only made sense for Kant if we presume that world exists for the sake of humans achieving their moral potential. Kant argues that for human beings who recognise this cosmic teleological vocation, this is sufficient for a subjective proof of the existence of God.[73] Again, similar to Richards's interpretation of Darwin, God was the selecting force behind natural selection. Yet Kant deviates from Richards's Darwin because Kant did not think that it was possible to provide theoretical justification for God's existence. Kant considers his account of teleological judgement to demonstrate the reality of a 'highest morally legislative author ... established merely for the practical use of our reason, without determining anything in regard to its existence theoretically'.[74] In other words, that we can think of God as the author of our moral action, and this enables us to subjectively assert that the purpose of the world is for us to achieve this end, does not determine anything about the world in terms of theoretical knowledge about the world.

For Richards, Kant and Whewell are not significant sources of influence on Darwin's theory. He argues that Kant's influence on Whewell 'forbade a philosophical leap into the transcendent sphere to explain the designed structure of organisms'.[75] Richards suggests that Whewell's position is similar to contemporary accounts of scientific creationism and intelligent design. Darwin revealed how we can combine theology and nature to understand nature as an intelligent selecting force. According to Richards, natural selection is an intelligent and altruistic selecting force located within nature: 'the process works for the good of the organism, unlike actions of the human breeder; that is, natural selection is an altruistic process, while human selection is selfish'.[76]

Artificial selection does not select on the basis of environments that are good for animals. Darwin considers cases where humans kept animals that are native to many different climates in the same conditions; providing them with the same exercise and feed, and often selecting half-monstrous forms to breed from. Artificial selection impedes natural selection, which only selects traits that make them better suited to their environments: 'natural selection can act only through and for the good of each being, yet characteristics and structures, which we are apt to consider as of very trifling importance, may thus be acted on'.[77] That natural selection is directed towards the good of each being is evidence for Darwin's argument that natural selection is an intentional progressive selecting force. Natural selection is a progressive force that selects traits resulting in improvements by making species better suited to their environment. According to Richards, '[n]o machine could see into the very fabric of creatures, could detect very small, virtually imperceptible, variations for selection'.[78]

Richards is correct to argue that Kant had little direct influence on Darwin, but again it seems there is partial overlap between their accounts. Kant's critical stance towards attempts to justify absolute purpose by appealing to relative purposes is analogous to Richards's identification of the tension between artificial and natural selection. Artificial selection creates machinations based on our limited scope of what is important, whereas natural selection selects for the good of the species. Natural selection could never be compared to a machine or artefact. Kant too denies that we can consider organisms as machines because they must be judged as possessing a formative force responsible for their generation, repair and reproduction. All these characteristics are absent in artefacts. My aim here is not to show that Kant did in fact influence Darwin, but to show that there is overlap between Kant's philosophy and certain aspects of interpretations of Darwin. Richards's emphasis on the role of God enables him to distance Darwin's account from a machine conception of the organism in a similar way to Kant's philosophy.

4.3 Organisms and design in contemporary biology

The role of teleology and design in biology have continued to be controversial issues for contemporary philosophy of biology. Kant's account of the organism offers an important perspective to these debates because he emphasises how our experience of organisms is dependent on judgement. Gould and Lewontin criticised the emphasis on adaptationist thinking within biology because it overlooked other factors that are significant for understanding biological development. This suggests that the way we understand biology is in part dependent on the conceptual framework that we apply to biological organisms. Following this, I examine Ratcliffe's argument that Dennett's account of adaptationism requires a non-naturalistic Kantian ground to explain how it is possible to view nature in accordance with intentions. Design and teleology have remained fundamental features of biological thinking.

4.3.1 Kant's account of design and its relation to contemporary philosophy of biology

For Kant, any organisms must be conceived as the 'cause and effect of itself',[79] which entails that the following three conditions are met: first, it possess the capacity for reproduction; second, it repairs and generates itself as an individual through growth; and third, its parts and whole are 'reciprocally dependent on the preservation of the other'.[80] Note that these features of entities judged in accordance with teleological principles do not relate to the specific traits that individuals have adapted in relation to their environment. Ginsborg refers to these features as the primitive conditions that any individual must fulfil to be judged as an organism. Her reference to these conditions as 'primitive' is central for her interpretation of Kant's account of teleological judgement. She argues that our judgements of organisms are primitive insofar as they are derived from neither theoretical nor practical reason; they relate to the conditions that any experience must fulfil to be judged as an organism.[81] This is different from discussions of teleology in contemporary biology, which are concerned with the use of specific terms such as *function* or *purpose*.[82] Kant argues that our justification for conceiving of nature as including these entities is not derived from nature, but from a teleological principle that cannot be derived from experience. In this sense, the conditions for the judgement of organisms is prior (or primitive) to the investigation of the particular traits of organisms. Kant identifies the tenuous relationship between design and biology, but contemporary biology regards design as merely an empirical problem. According to Ginsborg, the problem identified by Kant's conception of teleological judgement is fundamentally conceptual, and it relates to the requirement of viewing organisms *as if* they ought to possess certain traits:

> the empirical fact that an organism displays such-and-such a trait because that trait increased its ancestors' capacity to produce offspring, does not on its own entitle us to think of the animal as *designed* to have that trait, or more specifically, to claim that it *ought* to have it.[83]

Kant's account of teleological judgement gets to the heart of the controversy that persists in biology regarding the apparent design of biological phenomena. The predominant view in philosophy of biology is that biological functions are the product of natural selection, rather than being derived from psychological notions of purpose, design and intention.[84] Some biologists regard our use of design as a metaphor for explaining natural processes, yet design cannot simply be cast aside in biology once it has fulfilled its utility. For Ruse, without the metaphor of design, biology 'would grind to a halt'.[85] Evolutionary explanations are only possible on the basis that the capacity for the experience of organisms in general is pre-supposed; however, justification for this knowledge is beyond the remit of biology. They cannot 'say anything about the epistemological reasons that enable us to pick out something as a purposively organised unity in the first place ... with which the Kantian conception of teleology is concerned'.[86]

The tendency to consider biological traits in accordance with purposive design is central to the criticism of adaptationism by Gould and Lewontin. Adaptationism favours evolutionary explanations for the stabilisation of a trait within a species; a certain trait stabilises in a species because of the increase to fitness that it gives to those individuals possessing those traits. Gould and Lewontin argue that it is misleading for biology to investigate nature on the hypothesis that *all* traits stabilise because of increased fitness to the species. They argue that some traits stabilise because of physical constraints. These are analogous to architectural constraints such as spandrels. Spandrels refer to the spacers in buildings that are required when curved walls cannot meet. Gould and Lewontin justify this analogy between traits and spandrels since 'we find them [i.e., spandrels] easy to understand because we do not impose our biological biases upon them ... Since the spacers must exist, they are often used to ingenious ornamental effect'.[87] These spandrels help us to understand how the origin of a trait of an entity can be entirely different from its current use because the intricate decoration of these spandrels is not the primary reason for their existence.

Some biological traits may have originally stabilised because they were the by-product of other features of the organism. Like the spandrels, their current use may differ from the original reason for their existence: '[o]ne must not confuse the fact that a structure is used in some way ... with the primary evolutionary reason for its existence and confirmation'.[88] By explaining the stabilisation of traits primarily in relation to increased fitness, adaptationism tends to overlook alternative explanations. For instance, Dennett has argued that in cases where adaptationism cannot currently explain a trait, we should continue to look for adaptationist explanations: '[a]daptationist research always leaves unanswered questions for the next round'.[89]

Gould and Elizabeth Vrba argue that adaptation should be considered as one mode under the broader category aptation.[90] Aptation includes both adaptations and exaptations. Exaptations are further subdivided into cases where a trait is originally adapted for another function, or non-functional (or non-aptive) structures. An example of an exaptation is that the feathers of birds were primarily adapted for thermoregulation but were then utilised in the exaptive capacity for flight. Examples of exaptations that are non-aptive are more elusive. One reason for this is that they are a missing term in the taxonomy of evolutionary morphology, which means they do not feature in biological thinking. They argue that mutation at the genetic level has been accepted as non-aptive, '[b]ut we have not adequately appreciated that features of the phenotype themselves ... can also act as variants to enhance and restrict future evolutionary change'.[91] They argue that the current conviction of the supremacy of adaptationist explanations has resulted in an inability to comprehend the potential number of cases of exaptive traits. For these traits, 'current utility carries no automatic implication about historical origin'.[92] Gould and Vrba reveal that the adaptationist model cannot identify the historical origins of traits in all cases as it is blind to any other factors that contributed to the emergence of traits. The introduction of exaptations as a methodological term in biology allows us to consider how traits might not have stabilised in populations due to the fitness benefit of their current functions.

In contrast, Kant's account of the organism as the product of teleological judgement makes the stronger claim that it might not be possible to establish knowledge of the historical origins of organismic nature. Kant describes the investigation of nature from the perspective of historical origin as an archaeology of nature. It is 'a daring adventure of reason'.[93] This natural archaeologist assumes that the maternal womb of nature arose from chaos, but this does not avoid the need to regard nature as governed by teleological principles. According to Kant, 'he must attribute to this universal mother an organization purposively aimed at all these creatures, for otherwise the possibility of the purposive form of the products of the animal and vegetable kingdoms cannot be conceived at all'.[94] Kant pre-empted how the possibility of viewing nature in accordance with design required us to attribute nature with a principle of organisation.[95]

This provides a different perspective on the relation between Kant's account of teleology and scientific naturalism. The irreducibility of teleology and design within Kant's account is widely regarded to be a tension between Kant and biology. According to Zammito, '[t]he *third Critique* essentially proposed the reduction of biology to a kind of pre-scientific descriptivism, doomed *never* to attain authentic scientificity, to have its "Newton of the blade of grass"'.[96] In contrast to the idea that transcendental idealism is opposed to scientific naturalism and reduces science to a mere descriptivism, Ratcliffe has argued that Dennett's philosophy of biology is not as naturalist as it seems. Dennett's account of understanding design through the metaphor of 'reading Mother Nature's mind' cannot be supported by scientific naturalism. Ratcliffe argues that Dennett's account requires a transcendental non-naturalistic ground. For Dennett, natural selection requires us to view nature as if it were designed, and for us to regard nature as designed means we must regard it as the product of intentions. Dennett argues that intentionality disappears once we recognise that 'Mother Nature' can explain her selective procedure in terms of evolutionary processes, which are not intentional. Nature is designed without needing a designer. Dennett describes natural selection as 'a scheme for creating Design out of Chaos without the aid of Mind'.[97]

Dennett's account suggests that our understanding of design in nature requires this intentional stance, but this stance becomes unnecessary and should be removed once we possess knowledge of the non-intentional natural selective pressures that produce this design. Dennett does not recognise that it is impossible to explain away intentionality in this manner. According to Ratcliffe, '[t]he intentional stance cannot be eliminated, circumvented or explained away and, insofar as Dennett's account conceptually presupposes the intentional stance, it is ultimately nonnaturalistic'.[98] Far from showing that transcendental idealism is non-scientific because of its opposition to naturalism, it reveals that Dennett's account requires a non-naturalistic standpoint.

Ratcliffe suggests that we approach Dennett's account of intentionality from a Kantian perspective. For Kant, naturalism is an insufficient ground for us to judge nature in terms of final causes. Of course, Dennett is proposing that there is such a naturalistic basis; however, this presupposition stands in tension with his broader account. According to Ratcliffe, '[i]nstead of informing us about the structure of the world, he [i.e., Dennett] is inadvertently charting the constituting framework that renders a conception of the biological world possible'.[99] The possibility of conceiving of nature in accordance with design presupposes that we can conceive of the world *as if* it is designed.

For both Dennett and Ruse, the fundamental issue concerns the ability to adopt or dispense with the metaphor of design. Their conception of metaphorical thinking as a requirement for natural selection that can then be dispensed with once the non-intentional mechanisms have been uncovered is problematic. Dennett agrees that the original ability to perceive design in nature demands that we view nature in accordance with design, but his account suggests that this disappears once we gain a better understanding of the mechanisms of natural selection. For Kant, the original design cannot disappear from the conception of the organism without entailing the inability to recognise our experiences of organisms as such. According to Breitenbach, 'on Kant's account, to consider something in nature as organic is already to view it teleologically'.[100] If it is necessary that organisms are primarily conceived of in accordance with design, then it is also necessary to explain how the conception of the organism as the product of design can be replaced by a mechanical or non-teleological account.

4.3.2 Organisms without design

The distinction between man-made artefacts and organisms is still a pertinent topic in contemporary philosophy of biology. Nicholson identifies various disanalogies between organisms and artefacts that problematise the prevalent view of 'the machine conception of the organism'.[101] The fundamental differentiating feature between organisms and artefacts is that artefacts have functions at both the level of their parts and of their wholes, whereas organisms have functions only at the level of their parts. Taken as a whole, '[a]n organism does not have a function because its operation is not good for anything; it simply acts to ensure its continued existence'.[102] Nicholson notes that the domestication of animals is potentially problematic for this account, as domesticated organisms seem to display functions both at the level of their parts and their wholes. The organism's activity is hijacked by an external agent and subject to the purposes imposed on it by that external agent. It seems that, in such cases, domesticated organisms are also artefacts.

Elements of Kant's philosophy both support and critique the stance that organisms do not require any appeal to design. Kant would agree with the need to distinguish between organisms and machines. As previously discussed, machines cannot repair or reproduce their parts or their wholes.[103] Their functions depend on the intentions of an entity that is external to the machine. However, Kant argues that an organism must still be considered *as if* it were the product of intentional design, as this is the only way that we can regard the parts of the organism as existing for the sake of the whole. One fundamental difference between the accounts of the organism from Kant and Nicholson is that Nicholson argues that organisms are distinguished from artefacts ontologically rather than transcendentally, which is further justified by considering the emergence of organisms from the perspective of thermodynamics. Of specific importance is the discovery that all matter possesses a '*universal tendency* toward the degradation of mechanical energy'.[104] Unlike other entities, organisms are open to their environments, allowing them to maintain their organisation. Erwin Schrödinger describes this ability as '[a]n organism's astonishing gift of concentrating a "stream of order" on itself and thus escaping the decay into atomic chaos – of "drinking orderliness" from a suitable environment'.[105] Every part of an organism must replenish its orderliness from its environment, whereas, for machines, replenishment is specific to certain functions that they perform. According to Nicholson:

> This is why the fuel–food analogy is so misleading, and why the stability of a machine – despite its apparent dynamicity – ultimately resides in an unchanging material structure. In machines there is a specific 'inflow' and a specific 'outflow'. In organisms everything flows.[106]

It follows that conceiving of organisms as machines is misleading. Organisms are not designed in accordance with external plans,
nor is their energetic openness to environments limited to certain
functions. Organisms are better understood in relation to non-
organic dissipative structures that tend to maintain order, such as
storm systems or eddies. According to Dupré and Stephan Guttinger, '[t]he organism, thus broadly construed, can then be seen as
a stable eddy in the flow of interconnected biological processes'.[107]
Considering organisms as more complicated manifestations of
entities that do not require us to view them as the product of
design offers support to the argument that organisms are also not
designed. According to Nicholson, '[t]he striking thing about the
order of all dissipative structures, including organisms, is that it
arises in the absence of design'.[108]

Nicholson's account provides an explanation for why it is not
necessary to consider organisms as the product of design; however,
his account creates a difficulty for distinguishing between living
and non-living dissipative structures. Appeals to storm systems
and eddies highlight the way that the non-organismic systems also
possess clear boundaries that distinguish them from their environments without the need to appeal to design. Yet, there have been
some discussions about how to understand these non-organismic
dissipative structures. For instance, Kauffman has argued that the
storm system on Jupiter known as 'the Great Red Spot' could be
regarded as a living system:

> One can have a remarkably complex discussion about whether the
> Great Red Spot might be considered to be living – and if not, why not.
> After all, the Great Red Spot in some sense persists and adapts to its
> environment, shedding baby vortices as it does so.[109]

Considering organisms as non-dissipative structures entails that
organisms are not analogous to machines, which are the product of external intentions. Instead, they should be considered as
more complex manifestations of non-intentional, non-organismic
stabilisations of dissipative structures. On the one hand, this is
problematic for Kant's account of teleological judgement because it
reveals that organisms are not different in kind to non-organismic
dissipative structures. On the other hand, Kauffman's comment
that certain dissipative structures might be regarded as living can
be supported by Kant's account of teleological judgement.

Ginsborg has highlighted the virtue of Kant's account of teleological judgement is that it is applicable to non-organismic entities such as storm systems. She argues that philosophical analysis of the principle of function 'should not rule out in advance that we might encounter non-biological phenomena for which functional characterizations turn out to be scientifically indispensable'.[110] She specifies that entities such as storm systems are inherently non-biological because basic physical forces create them.[111] Yet, the ability to conceive of them in terms of teleological judgement suggests that it is not possible to establish this clear distinction between living and non-living dissipative structures, as we can judge both as organised systems in a way that differs from the kinds of functions that machines possess.

Some accounts of Kant have suggested if Kant had been aware of thermodynamics, then he would have radically altered his discussion of teleological judgement. They argue that Kant would have realised that it was not necessary to view organisms as governed in accordance with teleological principles. Thermodynamics provided the rule that allows for the manifest order of dissipative systems (or negentropic systems) to emerge from the broader context of global entropy. Alicia Roqué argues that Kant had to present mechanistic and teleological judgements of organisms as a regulative antinomy because he was not aware of entropy. According to Roqué, 'Kant regulates teleology to regulative judgment only because its peculiar form of recursive causality "unknown to us" could not be subsumed under a (mechanical) causal rule ... Nothing in Newtonian physics could provide such a rule'.[112] This sentiment is also reinforced by Andreas Weber and Francisco Varela when they assert that '[t]he real "Newton of the grassblade" was not to be an individual person, but a historical convergence of philosophical and biological thinking'.[113]

There are significant difficulties with the argument that discoveries in thermodynamics makes it possible to establish a compatibilism between the accounts of the organism conceived by Kant and contemporary philosophy of biology. Kant regarded our ability to engage in scientific enquiry as dependent on projecting a rational order onto nature in accordance with ideas. Scientific laws are not universal truths but regulative projections of order onto nature, hence it is not possible to provide a constitutive ontological account of the organism on the basis that we have discovered an appropriate constitutive ontological law of nature. Thermodynamics can only provide an ontological account of organisms under an inherently non-Kantian account of the relationship between epistemology and metaphysics. Under such an account, space and time are not limited to conditions of the subject, rather they are conditions of objects independent of experience. Moreover, Kant argued that we require teleological judgement to comprehend organisms in the first place, thus the thermodynamic account of dissipative structures does not refute the need to view them in accordance with teleological judgement to first comprehend them as stable self-organising systems.

Conclusion

In this chapter I examined the influence of Whewell and Herschel on Darwin's arguments from common ancestry and natural selection. The argument from common ancestry was more closely related to Whewell's conception of *vera causa* as derived from the principle of consilience, whereas Darwin's argument for natural selection drew from Herschel's argument from analogy by appealing to artificial selection. Darwin was attempting to satisfy the methodologies of both Herschel and Whewell.

There are significant theological implications for both arguments within Darwin's *Origin*. Whewell's account of consilience offered support to Darwin's argument that species originated from a single common ancestor. His account drew aspects of various disciplines together, including geography and embryology, yet he did not explain why consilience was so important for his theory. Whewell appealed to God to resolve the purposiveness of nature, the fundamental antithesis between thoughts and things, and explain how our scientific knowledge relates to truth. Likewise, the argument for natural selection presents nature as an all-powerful intentional selecting force analogous to our artificial selection of livestock to produce certain traits. Although there are disagreements about how to interpret natural selection, these do not mitigate the importance of theology for natural selection. God remains an important factor regardless of whether Darwin proposes natural selection as a machine conception of the organism, or as a non-mechanical intentional selecting force. Paley argued that organisms are evidence for divine design analogous to if we were to find a watch on a heath. Darwin did not question the analogy between the design of machines and organisms.

Kant's philosophy offers various criticisms against Darwin's account of the organism. Kant denied the analogy between organisms and machines. Ruse also appeals to Kant's regulative account of the organism as a way of understanding Darwin's account of organismic design. However, he misunderstands the role of regulative judgement within Kant's philosophy. We must judge organisms *as if* they are natural purposes not because we doubt that they are purposive, but because Kant could not reconcile this purposiveness with his broader scientific worldview.

Design remains an important issue for biology, Gould's and Lewontin's criticism of adaptationism highlights the biological assumption that all traits stabilise because they increase fitness for a species. In contrast, many traits stabilise because of physical constraints, which are analogous to architectural constraints called spacers or spandrels. The underlying issue is that the only possible explanation to explain the origin of a trait is not increased fitness. Moreover, the emphasis on adaptationism stands in tension with the biological commitment to naturalism. For instance, Ratcliffe argues that Dennett's account of adaptationism requires a non-naturalistic ground for his intentional stance. These appeals to Kant's philosophy expose how central principles of biology cannot be explained within the remit of naturalism.

5 • Kant's Significance for Contemporary Philosophy of Biology

Introduction

In this final chapter, I consider what a Kantian perspective can offer contemporary philosophy of biology in light of the argument that his philosophy has been influential for biology. The underlying issue is that appeals to Kant from contemporary philosophers of biology tend to isolate aspects of his critical philosophy from the broader context of transcendental idealism. In Chapter 1 I explained Zammito's criticism against appeals to Kant to solve current issues for philosophy of biology because of the broader incompatibility between transcendental idealism and naturalism. I argue that these appeals to Kant's philosophy need to be critically examined to appreciate the work that his philosophy is doing for these theories, and to identify the specific tensions that arise between the presupposition of biological naturalism and transcendental idealism. Not only does this reveal the points where these accounts have transgressed the limits of Kant's critical philosophy, but it also suggests how transcendental idealism could potentially offer support and direction for these theories.

The first section examines how definitions of biological individuality have remained a source of dispute in contemporary philosophy of biology. Differing conditions of biological identity offer incompatible accounts of the boundaries and number of organic entities. Generally, conceptions of biological individuality identify entities that express some degree of genetic homogeneity and functional integration. I examine paradigm cases of biological individuals as genetically homogeneous, symbiotic functional integrations and extended physiologies. The inability to agree on a single definition of an organism suggests that naturalism alone is insufficient for identifying the conditions for demarcating biological individuals.

The second section examines appeals to Kant's account of teleological judgement in support of contemporary accounts of biological autonomy. The Kant-inspired account of the organism, which emphasises the holistic conception of living systems, is undergoing a resurgence in contemporary biology. Appeals to Kant's philosophy within philosophy of biology are potentially misguided, as Kant's account of teleological judgement must be understood in the broader context of his critical philosophy.

I examine the relationship between physical teleology and moral teleology within Kant's account of teleological judgement. Kant's view was that biology was not the ground for freedom, rather freedom was the ground for conceiving of natural entities teleologically. This is the reverse of the contemporary conception of the relation between biology and freedom, which considers our freedom as dependent on our biological constitution. By drawing on prevalent interpretations of Kant's account of teleological judgement I will argue that, while this aspect of Kant's philosophy is often overlooked, it is of central importance. This importance is twofold. First, it exposes how Kant's account of physical teleology does not resolve its tenuous relation with naturalism. Second, it demonstrates how, by appealing to Kant's philosophy, this tension has been incorporated into aspects of contemporary philosophy of biology.

Finally, the third section considers how aspects of Kant's theory can offer guidance to contemporary philosophers of biology. Dupré has appealed to Kant's conception of moral freedom as possessing a degree of similarity with his own account of biological freedom. His argument is that the goal directed behaviours of humans have led to shared projects at the societal level that have increased the fecundity of individuals and species. These projects are overlooked under the assumption that all biological development can be explained at the genetic level. Both emphasise that humans should be considered as the source of their own agency. However, Dupré rejects Kant on the basis that he developed a deterministic account of nature. In contrast, I suggest that Kant develops a compatibilism between freedom and determinism. This compatibilism makes it possible for us to regard both the laws of nature and the laws of freedom as originating from the faculty of reason as regulative principles. Kant's philosophy potentially offers guidance for thinking about how our moral obligations related to politics. I consider how Kant could offer support to some of Dupre's implicit assumptions about how it is good to direct science towards investigations that will potentially result in societal benefits.

5.1 Definitions of biological individuality in contemporary philosophy of biology

The concept of the organism is central to biology, yet there is little consensus on what constitutes an organism or biological individual. Many accounts distinguish between technical and common-sense understandings of the organism. We all seem to have a common-sense awareness of organisms, yet, when this awareness is subject to further scrutiny, the distinction between living and non-living entities loses its clarity. According to Samir Okasha and Ellen Clarke:

> We know that dogs are organisms, while their tails are not ... Surely, then, any attempt to claim that species and organisms are problematic notions, calling out for the attention of philosophers, is just an example of professionals making work for themselves? The truth ... is that scientists really do encounter these problems, and the apparent obviousness of the intuitive concepts just makes our opening questions all the more pressing.[1]

Darwin assumed that the ability to identify organisms and species was common knowledge among naturalists and required no special treatment.[2] Developments in biology since Darwin have revealed examples of entities that appear to be organisms cannot be understood in terms of this common-sense account. For instance, when members from two separate species form a symbiotic mutualism that increases fecundity for both species, it can become difficult to identify the boundaries between these individuals, or to know how many individuals there are. These issues are important because 'if we cannot agree on the boundaries and number of individuals, we cannot obtain meaningful notions of populations. Without clear and non-controversial population structures, assessing the evolution of these systems is difficult at best'.[3]

There are various competing accounts for definitions of biological individuality in contemporary philosophy of biology.

Clarke identifies thirteen different definitive features that philosophers of biology have demarcated as conditions for biological
individuals.[4] John Pepper and Matthew Herron have supported the
co-existence of many concepts of the organism. Alternative organism concepts tend to refer to entities that express differing degrees
of functional integration and genetic homogeneity. There is a spectrum of definitions of what constitutes an organism that often
overlap, but each demarcates the biological kingdom in a different
way. There is not a categorical difference in kind between living and
non-living entities, but a continual difference in degree.[5] This section
examines a selection of cases that reveal the difficulties for establishing a single naturalistic explanation of biological individuality.

The Pando forest in Utah, United States, has been regarded as
the largest single living organism. Above the soil, each tree looks
separate from every other, yet, beneath the soil the trees all share a
common-root structure. These apparently separate quaking aspen
trees are clones formed by multiple runners fusing underground
and growing up towards the light.[6] The reasons for identifying the
Pando forest as a single biological individual is that it is genetically
homogenous and has distinct spatial boundaries, yet these boundaries are not visible above the soil.

In contrast, there are various examples of symbiotic relationships between genetically and spatially distinct organisms forming
relationships that increase the fitness for one or both sides of the
party. For instance, consider the symbiotic relationship between
the bobtail squid and the bacteria *Vibrio fischeri*. The squid provides protection for the bacteria by housing it in cavities in its
body (called photophores). The bacteria protect the squid against
predators as they possess the capacity for bioluminescence. The
predators of the squid hunt by looking for shadows that the squid
casts, but in this symbiotic relationship the squid is bioluminescent
and does not cast a shadow. The two organisms are genetically
distinct, yet functionally integrated in a way that increases fitness
for both parties. The functional integration of the squid and the
bacteria have been 'selected for' together as a whole via natural
selection. In this sense, the squid and bacteria could be regarded
as a single 'superorganism'. The trait that increases fitness is bioluminescence, which emerges only at the level of the symbiotic
relationship. The bacteria become bioluminescent through the
mechanism of quorum sensing. This means that when the bacteria

reach a certain density, the trait of bioluminescence activates. While the genes required for bioluminescence are located in the bacteria, the bacteria do not generally reach sufficiently high densities for bioluminescence outside of the squid's photophores. According to Frédéric Bouchard:

> if the bacteria alone does not often glow, and the squid alone cannot, who or what is bioluminescent and what benefits by being bioluminescent? Let us assume for a moment that we have an emergent trait (i.e. not reducible to single genome). What is the biological individual bearing that trait if not the system comprising both squid and bacteria?[7]

Bouchard considers the different ways that biological individuation could be applied to the symbionts. As there are 1 billion bacteria and 1 squid, we could say there are 1,000,000,001 biological individuals. Alternately, maybe there are two biological entities (1 squid and 1 Vibrio superorganism) or maybe there are 1,000,000,003 (1 squid, 1 billion Vibrio, 1 Vibrio superorganism, and 1 squid/colony emergent superorganism). In the context of this debate, what is important is the need for biologists to recognise the diverse ways that biological entities can respond to selective pressures – symbiotically or otherwise. These kinds of symbiotic relationship are not uncommon in nature, for instance, '[i]n contrast to the squid light organ, which is colonized with only a single symbiont, the mammalian intestine is inhabited by more than 400 species of bacteria. Similar to squid, mammals acquire their microflora from the environment'.[8] It has been estimated that there are ten times more microbes living inside and on our body than our somatic and germ cells.[9]

Some have suggested that our understanding of the organism should be extended to include the functional relationships that biological entities develop in relation to their environments.[10] For Scott Turner, the abiotic mound of a *Macrotermes* colony of termites ought to be considered as an extended organism. The mound is not biotic, but it 'is an organ of extended physiology that promotes the colony's respiratory gas exchange'.[11] It is necessary for the symbiotic relationship that the termites have formed with fungi that they bring into the mound to digest cellulose into simpler sugars that the termites then re-consume. Turner's argument is an extension of Richard Dawkins's account of the extended phenotype in the sense that an organism's behaviours and other activities are produced from the genes of that organism, and hence are part of that organism.[12] According to Turner, 'every organism is, in a sense, an extended organism, unable to exist without imparting a kind of physiology to its surroundings as well'.[13]

Both Turner and Bouchard support the idea that we can measure biological change at the genetic level, yet they oppose the notion that we could understand the potential functions made possible by genes by focusing solely on this level. It would not be possible to appreciate why the bobtail squid has the capacity to develop light organs unless we consider the symbiotic relationship between the squid and bacteria. According to Turner, 'genes really have no meaningful existence outside the functional environments that carry them into the future'.[14] Bouchard disagrees with the idea that these non-biological components of organisms could be understood in the same way as biotic aspects of an organism. He argues that '[i]f an evolving individual may be in part non-biological, then it is possible that its evolution will not be fully accounted for via differential reproductive success, since the adaptive external structures that Turner describes cannot be passed on genetically'.[15] There are many cases where extended physiologies could be utilised by subsequent generations. For example, a burrow can become part of the short-term extended physiology of an organism while it inhabits a burrow that it has constructed, but there are also cases where a burrow can become a persistent part of the functional environment 'where the same burrow was inhabited by many generations of burrow-diggers'.[16] The problem here is not whether the burrows or mounds have increased the fecundity of their dwellers, but whether these increases in fecundity can be accounted for genetically. The difference between the biological and non-biological components is that when non-biological components are 'inherited' by subsequent generations, the genes of the current generation have not directly contributed to the construction of this environment.

The idea that the ability to construct artefacts can be genetically inherited is relevant to the distinction between organisms and artefacts discussed in the previous chapter.[17] In this context, the issue is not the justification of organisms as analogous to artefacts, but rather whether the artefacts that organisms produce can be conceived of as part of the organism. John Symons has suggested that artefacts and organisms are ontologically distinct. According to Symons, '[t]he factor that determines the persistence condition of the organism is the functional interdependence ... and not the origin of that interdependence'.[18] This distinction between organisms and artefacts does not relate to the design or plan of an artefact as external to the organism as it did for Kant and Paley.[19] For Symons, both genetically engineered organisms and organisms that are artificially selected in virtue of a trait for our purposes cannot be considered as artefacts. These changes are introduced artificially, but the changes are functionally interdependent and cannot be separated from the organism. In contrast, a pacemaker is an artefact because it does not depend on the other functions of the body. When an individual with a pacemaker dies, 'the pacemaker will still be a pacemaker and could possibly even be reused in another body'.[20]

According to Symons, entities such as termite mounds and burrows would also be considered artefacts insofar as they do not become functionally integrated with organisms. They can exist independently from the organisms that first created them and can be inhabited by future generations, which is analogous to the example of the pacemaker that could be transplanted to another body. Hence, Symons's distinction between artefacts and organisms offers support to the argument that the termite mound could not be considered as an organism because it is not completely functionally integrated with the termite colony. However, the *Vibrio fischeri* that are housed in the light organ of the bobtail squid would also be understood as artefacts for the same reason. The squid expels most of the bacteria daily, so they also do not become fully integrated with the squid. According to Bouchard, this is 'most likely to reduce the possibility that the symbiont would evolve a pathogenic response as many other Vibrio have done'.[21]

These examples demonstrate that the ability to offer a single definition of the organism is far from clear. This has led to doubts regarding the traditional conception of the organism as a paradigm individual. Bouchard has criticised Peter Van Inwagen's metaphysical conception of organisms as possessing a special causal relationship that the biologist is responsible for explaining. Bouchard argues that Van Inwagen is sidestepping the issue of vitalism for biological composite entities 'by putting in the hands of biologists the problem of figuring out the nature of this special causal relationship'.[22] Against Van Inwagen, biologists have not generally found this special causal relationship. Instead, biologists have found that organisms are generally not paradigm cases of this form of individuality at all.

In summary, this section has surveyed some difficulties surrounding definitions of biological individuals in contemporary philosophy of biology. These difficulties arise when the notion of biological individuality is regarded as derived entirely from nature without recognising the importance of our faculties for comprehending these entities as organisms in the first place. How we understand biological individuality depends on the emphasis placed on factors such as genetic homogeneity or functional integration. Genetic homogeneity can explain why we judge the Pando forest of quaking aspen as a single biological entity, but genetic homogeneity cannot sufficiently explain how we judge symbiotic relationships as biological individuals. In cases such as the symbiotic relationship between the bobtail squid and *Vibrio fischeri*, the squid has developed light organs in response to the presence of the bacteria. When entities with distinct genetic lineages become functionally integrated with one another, this also encourages us to judge the experience as representing a single biological entity. Turner expands this notion by arguing that our understanding of functional integration should be extended to abiotic structures that organisms produce. The abiotic mounds created by termite colonies should be regarded as an extended physiology of the organism. The mound building behaviour of the colony shows how the colony possesses the genes to turn their environment into an extended physiology. While all three cases emphasise the importance of genetics for understanding biological individuals, disagreement emerges between their alternative explanations of the boundaries of these individuals. The inability for biologists to reach a consensus on a definition of biological individuality offers additional support to the Kantian idea that organisms are inseparable from the way that we judge them.

5.2 Natural teleology and biological autonomy

The contemporary account of biological autonomy draws explicitly from Kant's account of physical teleology. Kant's account of teleological judgement offers support to their notions of biological closure and Kantian wholes, which are central concepts associated with biological autonomy. In contrast, I argue that when Kant's conception of teleological judgement is understood in the context of transcendental idealism and, specifically, its relation to his account of moral teleology, it does not offer support to the position of biological autonomy. Kant's focus on moral teleology reveals that his primary concern is not directed towards explaining the emergence of natural entities that possess the capacity for self-organisation. Rather, Kant is concerned with how our ability to *judge* nature as possessing these kinds of entities is evidence for the existence of freedom. I argue that these two kinds of teleology – moral and physical – should be interpreted as analogical counterparts of the power of reason to view nature and freedom in accordance with ends. Kant's argument that physical teleology depends on moral teleology is often overlooked by contemporary Kantian scholarship.

5.2.1 Biological autonomy in contemporary philosophy of biology

Kant's conception of the organism has been influential for various accounts of agency in contemporary philosophy of biology. For instance, Kauffman appeals to the conception of Kantian wholes as a way of understanding organisms as possessing their own agency and autonomy in their process of self-recreation. He argues that Kant's account helps us to understand the essential relationship between organisms and functions. According to Kauffman:

> It is this combination of self-recreation of a Kantian whole, and therefore, its very existence in the non-ergodic universe above the level of atoms that, I claim, fully legitimizes the word, 'function' of a part of a whole in an organism.[23]

Kauffman deploys Kant's account of organisms in the third *Critique* to explain how organisms must first possess the ability to produce a function that can become a unit of selection. The fundamental capacities that enable organisms to produce functions are self-reproduction and metabolism. Kauffman defines metabolism as a process in which the activity or work carries out a work cycle.[24]

Work cycles explain how both organic and non-organic systems utilise energy from their environments to perform an activity that results in the final organisation of the relevant parts and whole of the entity as ready for another cycle of work. Work cycles cannot occur at equilibrium; they require a differential arrangement of parts to make it possible for an entity to perform a function.

Work cycles can be applied to living and non-living systems, but the difference is that living systems establish their own boundaries and causal closure, whereas for non-living systems these features are constructed externally. The difference between living and non-living systems is also central to the account of biological autonomy developed by Matteo Mossio and Alvaro Moreno. They argue that the paradigm example of a biologically autonomous individual is a genetically homogenous multicellular organism. One distinguishing feature between organisms and artefacts is that organisms 'result from a process of *differentiation* between their functional parts, and not from the *aggregation* of pre-existing entities'.[25] They argue that this process of differentiation of functional parts emerges internally from within the organism and appeal to Kant's distinction between organisms and artefacts to support their account. According to Mossio and Moreno:

> organisms realise closure, while artefacts do not. Accordingly, while for the Darwinian tradition, the comparison between a watch and an organism – even regarding only contrivance and relation between parts – suggests an analogy, the organisational view requires an essential distinction.[26]

Kauffman also appeals to Kant's account of the organism to describe how cells carry out constraint construction and other construction projects (e.g., DNA replication), establishing a closure that enables the cell to create a rough copy of itself. According to Kauffman, 'this whole process is precisely the self-propagating organization to which Kant pointed'.[27] From this, Kauffman argues that life can be understood as an ontologically emergent process that involves a reciprocal relationship between matter, energy, information, entropy, and more. For Kauffman, these Kantian wholes are a prerequisite for natural selection as natural selection can only act on entities that possess the capacity for self-propagating organisation.

Mossio and Moreno describe the causal cycle of constraints through the following thought experiment. Imagine there are two billiard balls on a table and the table is configured in such a way that when the first ball hits the second ball, the second ball comes back and hits the first ball, starting the cycle over again. In ideal frictionless conditions, the constraints (table, boarders, balls) will entail that the initial application of a force will result in 'a causal cycle of constrained processes'.[28] In this situation, a causal cycle is established, but not organisational closure. Organisational closure refers to cases where a system generates its own constraints, but in this thought experiment the constraints are externally imposed. Biological closure requires the entity to possess sufficient internal complexity to take control of various environmental factors that contribute to the boundary conditions required for the maintenance of the system. Biological closure achieves a greater level of autonomy and independence from perturbations in environments.

In contrast, non-biological self-organising systems (or dissipative structures) contribute to their own maintenance only in a single way and small variations in the environment can result in the loss of organisation for the whole structure. Biologically autonomous systems achieve biological closure because they possess a physical boundary that non-living self-organising systems lack. According to Moreno and Mossio, '[w]hile self-organising systems are not delimited by a physical border, all biological cells possess a membrane ... Membranes help distinguish the system from the environment, while at the same time enabling it to act on relevant factors'.[29] The difference between non-living self-organising systems and biological systems is a difference in kind rather than degree. They argue that it is not possible to understand the autonomy of biological systems by looking solely at 'self-organisation as it occurs in physics and chemistry'.[30]

Their separation of living systems from more basic self-organising systems is in tension with their claim that life must have originally emerged from non-living dissipative structures as both are ultimately grounded in thermodynamics. Nicholson argues that it is not possible to discount the importance of non-living dissipative structures when considering the development of autonomous biological systems.[31] One reason why Mossio and Moreno might insist on such a strong distinction between these kinds of entity is because of their strong Kantian leaning. They argue that when causal closure is established, the emergent entity possesses the irreducible capacity to self-determine as a whole.[32] In contrast to Kant, they argue that this does not require that the whole possesses a distinctive causal power, as the network of interrelated causal parts is the source of the power of the whole.[33] Nevertheless, their alignment to Kant's philosophy requires a closer examination of Kant's account to appreciate the context of the emergence of Kant's discussion of physical teleology and the issues that it was intended to resolve.

5.2.2 The relationship between physical and moral teleology in Kant's philosophy

Kant argued that our ability to judge an entity as if it is governed by natural purpose requires both theoretical and practical reason. A natural purpose draws from practical reason insofar as the parts and whole act reciprocally for the sake of one another. In contrast, for efficient causation, the whole can neither precede the parts nor be the purpose that the parts exist for. A natural purpose is subject to temporal conditions insofar as it is an object of experience. Kant argues that space and time are necessary conditions for the possibility of any experience. A natural purpose is also subject to the determinations of efficient causation and therefore cannot be explained in the way that moral maxims of practical reason are (i.e., categorical imperatives). Kant did not think that teleological judgements could reveal any knowledge of freedom, rather he asserted that physical teleology needed to borrow from principles of moral teleology. According to Kant:

> **moral** teleology makes good the defect of **physical** teleology, and first establishes a **theology**; since the latter, if it is to proceed consistently rather than borrowing, unnoticed, from the former, could by itself alone establish nothing more than a **demonology**, which is not capable of any determinate concept.[34]

When physical teleology is taken in isolation from moral teleology, it could lead us to regard the purposiveness of biological organisms as justification for the existence of God, which would be a demonology. Instead, our ability to conceive of physical teleology is made possible by our capacity for practical reason: 'given the subjective constitution of our reason … this final end can be nothing other than **the human being under moral laws,** while by contrast the ends of nature in the physical order cannot be cognized *a priori* at all'.[35] In other words, as it is not possible to comprehend the ends of nature as anything other than humanity, we should look to our moral ends as a ground for physical teleology.

The idea that moral teleology grounds physical teleology is not a widely held view in Kant scholarship. Ginsborg argues that our ability to judge experience in accordance with physical teleology is not associated with the imperatives of practical reason. For Ginsborg, the difference between the practical and physical conceptions of teleology is that practical teleology ascribes a value as a maxim, such as a categorical imperative, which is deemed to be good or rationally desirable. In contrast, judgements pertaining to physical teleology are not developed in accordance with a maxim and therefore cannot be regarded as rationally desirable or good in the sense of moral teleology.[36] In the absence of rational justification for physical teleology, Ginsborg argues that the entities judged as displaying physical teleology manifest their own conditions for their primitive normativity. This means that when we perceive an organism such as a horseshoe crab and judge that it ought to have eight legs, then 'our judging that they are as they ought to be is a condition of our judging that horseshoe crabs ought to have eight legs in the first place'.[37] The difference between living and non-living entities is that only living entities can be judged in accordance with normative constraints. For instance, there is no way that a stone ought to be. Importantly, this is not a moral sense of ought; horseshoe crabs that have fewer legs, or eyes that lack the capacity for vision, are defective in the sense that they do not meet the criterion for their primitive normativity, but they are not morally defective.

Ginsborg is correct to separate the physical and moral notions of teleology to explain how we could not infer that something was morally defective from judging an entity as physically teleological. Kant distinguished these two aspects: 'such dissimilar principles as nature and freedom can only yield two different kinds of proof.'[38] However, there is an alternative way for moral and physical teleology to offer support to one another. The ability to conceive of both nature and freedom in accordance with ends supports the human propensity towards the power of reason. In other words, physical teleology is supported by moral teleology because it is an analogical case in which nature can be considered in accordance with final causes. Moreover, when physical and moral teleology are taken in combination, this reveals that nature cannot be considered as the source of this capacity as only moral teleology can be grounded *a priori* in our capacity for practical reasoning.

Kant's argument that we conceive of natural purposes by analogy with our own rational capacities has been a subject of dispute. For instance, our ability to conceive of organisms in accordance with our own rational capacities has been interpreted by Breitenbach as extending beyond our 'ability for free and end-directed activity ... presenting a complex capacity whose different functions are purposively related to realizing and maintaining the capacity of reason as a whole'.[39] Against Breitenbach and Ginsborg, I argue that the analogy with our own rational capacities should be understood as relating specifically to the relationship between moral and physical teleology.

There is textual support for my interpretation from the concluding section of the third *Critique*, which offers a comparison of the relationship between moral teleology and the physico-theological proof of God's existence. The physico-theological proof suggests that God's existence can be derived from the intentional design of nature. Kant argues that the physico-theological proof cannot sufficiently justify God's existence because this would require us to conceive of the intelligence of this original being.[40] It is impossible to conceive of such an intelligence because it would demand the transcendence of our own intelligence to view an intelligence greater than our own. Instead of proving the existence of God, moral teleology supplements the shortcomings of physical teleology because it 'rests on *a priori* principles that are inseparable for our reason'.[41] Moral teleology offers support to physical teleology, but physical teleology does not offer substantive support for moral teleology. According to Kant, '[t]he moral proof ... would thus always remain in force even if we found in the world no material for physical teleology at all'.[42]

The emphasis of Kant's argument is that our capacity to view nature teleologically is grounded in our capacity for practical reason and morality. Natural teleology merely offers corroboration – but not proof – for moral teleology. Moral teleology is grounded in *a priori* principles that are inseparable from reason. Importantly, physical teleology cannot offer confirmation for the conclusion of the moral proof, which is directed towards demonstrating that God is indispensable for practical reason. Rather, it supports moral teleology because nature is 'capable of displaying something analogous to the (moral) ideas of reason'.[43]

My interpretation highlights an aspect of Kant's philosophy that is overlooked in contemporary Kant scholarship, namely one of Kant's intentions for the third *Critique* was to rectify the great chasm between theoretical and practical reason that had emerged from the first and second *Critiques*.

The issue with the interpretations offered by Ginsborg and Breitenbach is that they both concede too much to nature by overlooking the role of practical reason and the supersensible for teleological judgement. In response to Ginsborg, our ability to judge entities as displaying the conditions of their own primitive normativity requires that we already view nature in accordance with a concept of causality through freedom. In response to Breitenbach, our different capacities that realise and maintain reason as a whole must also ultimately be related to this causality through freedom. The problem for both is that it is not possible to derive our concept of freedom required for physical teleology from nature without appealing to a supersensible notion of freedom associated with Kant's account of practical reason. Of course, this kind of freedom can only be known as a regulative principle for two reasons. First, while moral principles are necessary for practical reason, they are merely contingent from the perspective of nature and theoretical reason. Theoretical reason is descriptive rather than prescriptive; although moral maxims *should* be followed, they often are not. Second, physical teleology only corroborates moral teleology insofar as it reveals an analogous manifestation of our ability to judge in accordance with ends.

Kant's appeal to a supersensible conception of freedom is of particular importance when considered in the context of appeals to Kant's philosophy by contemporary philosophers of biology. There are significant differences between Kant's account of the relationship between physical and moral teleology and the conception of Kantian wholes appealed to by advocates of biological autonomy. Moral teleology does not require that nature display teleology, but the fact that it does offers additional corroboration for our capacity for moral reasoning. We can infer from this that it would not be possible to conceive of nature teleologically if we did not have the capacity for moral teleology. The contemporary account of biological autonomy borrows from Kant's account of physical teleology to explain organisms as biologically autonomous wholes, which are responsible for their own self-organisation. Kant's critical philosophy does not offer sufficient justification for the conclusion that nature contains entities that possess the capacity for self-organisation, what Kauffman terms 'Kantian wholes'. For Kant, our ability to perceive such entities is dependent on our moral reason, but this still relates only to our ability to *judge* nature as possessing such entities rather than revealing that such entities persist independent of experience.

5.3 Kantian morality and biological freedom

Within biology specific interest has been focused on the development of humans: '[t]he human individual is idiosyncratically worthy of respect and moral treatment.'[44] The idea that humans are in some sense unique because of their idiosyncratic worth is based on their capacity for autonomy and freedom, yet these features do not fit well with some of the naturalistic presuppositions of the biological sciences. Peter Godfrey-Smith considers the difference between moral and legal individuality. He suggests that we separate the legal and biological senses of individuality: '[m]aybe we should say different things in different contexts. Monozygotic human twins deserve two votes in elections, but perhaps they form a single unity in another sense'.[45] The idea that we should distinguish between the biological and legal senses of individuality resonates with Kant's philosophy. In his political philosophy, Kant is concerned with how we ascribe and respect the rights of one another. Our status as legal entities possessing rights goes beyond the remit of the naturalist explanations of biology. Although our sense of political individuality is distinct from biological individuality, the actions and decisions made at the political level have important ramifications for understanding biological change. Kant's philosophy offers a way to understand how a more complete account of human practices needs to consider the interrelation between our different activities. Science and society are inseparably connected;

science is not value free. Kant's political philosophy provides a framework for understanding broader implications and responsibilities relating to scientific research.

5.3.1 Dupré's account of biological freedom

Dupré has considered how his account of freedom as emerging from biological processes is similar to Kant's conception of freedom. He is careful to avoid any possible implication that would commit his account to the broader metaphysical tenets of Kant's philosophy. In this sense, it would be misleading to suggest that Kant had influenced Dupré's conception of biological autonomy; rather, there is a similarity between Kant and Dupré in regard to their conceptions of agency. According to Dupré, 'Kant's account of human action points in the right direction in which to look for this final ingredient of an account of human autonomy'.[46] Kant's philosophy is in the right direction because it is amenable to showing how humans can direct their activities towards their own goals. The impact that these projects have on our lives cannot be understood if we limit the focus of biology to genetic factors alone. We also need to consider how organisms actively construct their environments (or niches) in ways that promote conditions for survival and development that are not reducible to the genetic level. Dupré considers the emergence of hospitals and schools as examples of entities emerging from goals at societal levels that have contributed to both the extension of our lives and potential skills we can learn. We remain blind to the biological impacts of such entities under a reductive genetic biological account.[47]

Dupré distinguishes his account from Kant's on the basis that his conception of autonomy is not a 'rationally grounded canon of morality that constitutes an action as free or unfree'.[48] Instead, he argues that the difference between free and unfree actions should be understood in terms of 'a spectrum of degrees of causal efficacy'.[49] In contrast to Kant, this spectrum of causal efficacy does not require an alternative kind of causality to account for freedom in opposition to the causality that governs nature more generally. For Dupré, the rejection of determinism at the ontological level entails that the causal capacities at the level of human activity are the 'densest concentration of causal capacities, or causal powers, in our experience'.[50]

The difference between the accounts of freedom developed by Kant and Dupré is that Kant denied any possibility of experiencing freedom as a spatio-temporal phenomenon. However, this does not mean that Kant was a determinist. He argued that the possibility of freedom was established by revealing the apparent determination of the world related only to the level of appearance, rather than determining the world as it is independent of experience. According to Kant, '[a]ll manifoldness of things is only so many different ways of limiting the concept of the highest reality ... [T]his does not signify the objective relation of an actual object to other things, but only that of an **idea** to **concepts**'.[51] Transcendental idealism entailed that, at the level of ideas, ultimate enquiry into the ontological reality remains in a state of indeterminacy, rather than relating to objects in themselves. In the context of Kant's theoretical philosophy, he developed an account of causal determinism at the level of the appearance of objects, but he denied that this determinism at the level of appearances culminates in a constitutive understanding of reality as a totality. According to Kant, for 'the **concept** of all reality, reason only grounded the thoroughgoing determination of things in general, without demanding that this reality should be given objectively, and itself constitute a thing'.[52]

5.3.2 Determinism and the Second Analogy

To show how Kant's account avoids ontological determinism, it is helpful to consider his argument in more detail. Kant explains how experience must be understood as governed by efficient causation in the Second Analogy. In cases of efficient causation, the cause always precedes the effect. Efficient causation must be understood in terms of the temporal distinction between the cause and the effect. Kant distinguishes between objective and subjective experiences of succession by contrasting the experience of seeing a ship going down a stream with viewing a house. In the former case, the appearance of the ship going down a stream requires that there was a previous perception of the ship further upstream. According to Kant, 'it is impossible that in the apprehension of this appearance the ship should first be perceived downstream and afterwards upstream'.[53] The experience is objective insofar as we can say something about the object of experience; this ship has gone downstream. The sequence or order of the experience is essential to the concepts that we use to understand the experience. Objective succession requires that we can establish a rule that determines experience in relation to the temporal order of appearances. If the ordering of the experience were reversed, then it would refer to different state of affairs; the ship would have gone upstream.

In contrast, the experience of viewing a house does not depend on any specific ordering of successive appearances as we can view the different features of the house in multiple ways. Although the experience of the house is successive, it is not possible to discover a rule that determines the order we must view the different parts of the house. No matter how we view the house, our experience of the house does not reveal to us any change in the object of appearance. To be clear, the necessity that Kant attaches to objective succession does not pertain to the lapse in time, but to its temporal ordering. Kant argues that even in cases of efficient causation where the cause is simultaneous with the effect, such as a stove heating a room or a ball making a dent on a pillow, we are still able to derive the rule that makes it possible to view such experiences as objective. According to Kant, 'it is the **order** of time and not its **lapse** that is taken account of; the relation remains even if no time has elapsed'.[54]

Dupré characterises Kant as advocating a deterministic philosophy that 'could be deduced *a priori* from the possibility of knowledge'.[55] In the context of the Second Analogy, this characterisation is correct. Moreover, Kant argues that the role of the understanding in relation to nature is not restricted to merely establishing rules through the comparison of appearances, rather 'it is itself the legislation for nature, i.e. without the understanding there would be no nature at all'.[56] The connection between the faculty of understanding and the laws of nature does not extend the scope of the understanding to knowledge of the laws of nature independent of appearances. Rather, Kant reduces knowledge of nature to appearances. There are two different senses that the laws of nature can be understood in Kant's philosophy. First, laws corresponding to the pure categories that are necessary for the possibility of experience. Second, general scientific laws of nature relating to regulative ideas that do not directly correspond to objects of experience, rather they are principles that are projected onto nature by reason as guidelines that direct enquiry.

In Chapter 3 I explained how Kant describes ideas as guiding principles for engaging in scientific understanding through his discussion of the *focus imaginarius*, which appealed to Newton's example in the *Opticks*.[57] The *focus imaginarius* is an optical illusion whereby an image in a mirror is believed to relate to an object behind the mirror. In the context of transcendental idealism, the aim of the example was to demonstrate the necessary illusion that arises when concepts extend indeterminately in relation to the guidance of the unconditioned regulative principles of reason.[58] It is natural for us to assume that these principles uncover knowledge about an object independent of experience in an analogous manner to the assumption that there in an object behind the mirror, but this is a mistake as the object is a product of the illusion. The only way to avoid this illusion resulting in deception is to expose how these projections are not 'shot out from an object lying outside the field of possible empirical cognition'.[59] For Kant, the systematicity that guides experience does not relate to objects in themselves, but rather it emerges from the demand of reason to impose the greatest possible unity on experience.

Despite Dupré's critical stance towards Kant because of his commitment to determinism, there is a degree of compatibility between their accounts, as both Kant and Dupré reject determinism at the ontological level. Kant's account of causality is specifically focused on the necessary temporal order of certain concepts, rather than discovering necessary relations between objects in themselves.

5.3.3 The role of Kant's account of freedom for guiding the development of science

Kant's demarcation of things in themselves from objects of appearance preserved the possibility of freedom. According to Kant, 'if appearances are things in themselves, then freedom cannot be saved'.[60] In the context of practical reason, we must consider ourselves as governed by an entirely different kind of causality. The account of causal necessity that Kant developed in his theoretical philosophy referred only to efficient causation under the conditions of appearances. This is not applicable to practical reason because practical reason is not limited to the conditions of appearances. According to Kant:

> natural necessity ... attaches merely to the determinations of a thing which stands under conditions of time and so only to the determinations of the acting subject as appearance ... But the very same subject, being on the other side conscious of himself as a thing in itself, also views his existence *insofar as it does not stand under conditions of time* and himself as determinable only through laws that he gives himself by reason.[61]

The relation between practical reason and time was previously discussed in the context of the two-world and two-aspect interpretations of Kant.[62] I argued that Allison's two-aspect interpretation separated causes from reasons on the basis that actions governed by practical reason can never be understood as a proof of freedom, but rather freedom must be presupposed to understand how it is possible to formulate rational maxims to guide enquiry. In contrast, the two-world interpretation argues that freedom constitutes another kind of causality directed towards ourselves as noumenal agents. Clearly, Kant does consider freedom to relate to a different kind of causality. According to Kant, '[t]he concept of causality as *natural necessity*, as distinguished from the concept of causality as *freedom*, concerns only the existence of things insofar as it is *determinable in time*, and hence as appearances, as opposed to their causality as things in themselves'.[63]

Wood describes the compatibility between freedom and nature in Kant's philosophy as 'the compatibility of compatibilism and incompatibilism'.[64] These two forms of reasoning are compatible so long as we understand that one cannot encroach on the domain of the other. The essential difference is that freedom is entirely separated from intuitions. Kant argues that the entities that are central to practical reason (freedom, immortality and the soul) are pure rational concepts: 'no corresponding intuition can be found and consequently, by the theoretical path, no objective reality.'[65] As these objects are not given in intuition, they are necessarily non-spatio-temporal. Thus, the grounds for claiming a compatibility between freedom and nature resides in Kant's argument that the determinism of nature relates only to objects of appearance and not to objects as they are in themselves.

This helps us to understand why Kant developed his account of freedom as entirely distinct from the temporal conditions that determined entities understood through theoretical reason. When we act in accordance with our inclinations, we surrender our autonomy to past events, which reduce our agency to nothing. In other words, we subject ourselves to the determination of efficient causation because the reasons for our actions can be understood only as efficiently determined by previous temporal events. This is consistent with Kant's distinction between hypothetical and categorical imperatives. Hypothetical imperatives require that the means and ends of an action are temporally distinct. According to Kant:

> all imperatives command either *hypothetically* or *categorically*. The former represent the practical necessity of possible action as a means to achieving something else that one wills ... The categorical imperative would be that which represented an action as objectively necessary of itself.[66]

The categorical imperative makes it possible to formulate maxims that govern our actions where the end of an action is not distinct from the means to achieve that end. Categorical imperatives ensure that the motivation for our action is our rational sense of duty, rather than a particular inclination or incentive that might motivate us to act at that particular moment.

Kant's distinction between theoretical and practical reason is indicative of deeper tensions between transcendental idealism and naturalism. For instance, many explanations for the emergence of morality in contemporary philosophy of biology presuppose that the emergence of morality must have a naturalistic explanation. David Hull associates the idea of a specifically human nature with ethics and morality. The concept of legal right is premised on the notion that we have a high degree of similarity with one another. Hull argues that this assumption is inconsistent with our current biological understanding: '[a]ny ethical system that depends on all people being essentially the same is mistaken. If we have rights, we must have rights even if we are not all basically the same'.[67] The theory of natural selection depends upon diversity and variability both within and between species. Biology reveals to us that all organisms and traits are contingent. If there is a universal human nature, then it is grounded on the principle of variation that humans share with all biological species. According to Hull, '[w]hich variations characterize a particular species is to a large extent accidental; *that* variation characterizes species as such is not'.[68]

Hull correctly identifies a tension between Kantian morality and biology. Kant's separation of freedom from experience
requires that we view the source of free action as outside of time,
and thus outside of nature. For Kant, moral agents must consider
themselves as 'a lawgiving member of the universal kingdom of
ends'.[69] Rational agents are homogenous with one another insofar
as they possess autonomy. Kant sets 'the principle of the **autonomy** of the will in contrast with every other, which I accordingly
count as **heteronomy**'.[70] Like Hull, he was also aware that empiricism could only result in knowledge of an essentially contingent
kind in relation to morality. Hull was merely cautious of deriving moral principles from contingent conditions, whereas Kant
denies any such possibility. Practical reason derives 'its concepts
and laws from pure reason, to set them forth pure and unmixed,
and indeed to determine the extent of this entire practical or pure
rational cognition'.[71] When the faculty of reason is not the source
of an action, the action cannot be considered as moral. For any
finite rational being, action can be caused by either the faculty of
desire or the faculty of reason. We can be inclined towards actions
that correspond to a need. However, if we act upon this inclination, then the cause of that action is necessarily a contingent
state of affairs that originates from experience, not from reason.
An action caused by inclination is heterogeneous and contingent
as it is caused by something outside of itself. It does not signify a
moral action but rather a '*pathological* interest in the object of the
action'.[72] Moral actions are antipodal to actions based on inclinations and incentives. Action-based inclinations relate to the values
that we ascribe to experiential objects, which are the product of
schematising intuitions and categories together. In contrast, practical reason has no relation to experience as the maxims that form
categorical imperatives have no relation to intuitions.[73] Hull's commitment to biological naturalism entails that he cannot conceive
of any appropriate grounds for concepts such as homogeneity and
autonomy within the remit of naturalism. Hull is left with the following conclusion:

> Although I feel uneasy about founding something as important as ethics
> and morality on evolutionary contingencies, I must admit that none of
> the other foundations suggested for morality provides much in the way
> of a legitimate sense of security either.[74]

There is an important distinction between the accounts of Kant and Hull that relates to the idea of the unity of science (or consilience) discussed in Chapter 3.[75] The idea that biology will offer a naturalistic explanation for the emergence of morality, despite lacking evidence to support this claim, demonstrates the level of enthusiasm and hope that biologists have for the scope and unifying power of biological explanations.[76] In contrast, Kant's account of reason as determining, but not itself determinable, makes it possible for Kant to separate the practical and theoretical aspects of reason in a manner that does not result in the need to establish a single reductive principle by which all human activity must be ultimately explainable. Kant's account of the regulative unity of science separates the different activities of humans (or finite rational beings) and explains how these activities can all be justified in accordance with the architectonic structure of transcendental idealism. He emphasises that unity in nature is the product of the demand of reason. Ideas are not determined by nature in either their theoretical or practical applications. In relation to theoretical reason:

> One cannot properly say that this idea is the concept of an object, but only that of the thoroughgoing unity of these concepts, insofar as the idea serves the understanding as a rule. Such concepts of reason are not created by nature, rather we question nature according to these ideas.[77]

Reason in its theoretical use only determines appearances, not objects in themselves. It provides the understanding with a guideline by which it can search for increasing unity in its determinate use, while simultaneously recognising that such unity will never be sufficiently demonstrated for nature as a totality, rather, it must be presupposed. For Kant, this role of reason is heuristic, but not in the sense that it is merely one possible way that we can increase the scope of nature by appealing to an unconditioned principle. It is essential for the possibility of empirical enquiry; without reason we would have 'no coherent use of the understanding, and, lacking that, no sufficient mark of empirical truth'.[78]

The indeterminacy of the idea of nature as a totality relates to the previous discussion of Breitenbach's and Choi's (2017) account of unified pluralism discussed in Chapter 3.[79] They emphasised the epistemic virtue of the Kantian account is that it demonstrates the need for philosophers of science to approach nature as if it were unified in order to establish science as a communicative and collaborative process. They argued that such a unity does not ever need to be ultimately discovered, although this remains an open possibility. On my interpretation of Kant, this unity could not be discovered in principle, as this would require the collapse of the 'gap' between the undetermined idea of unity and the determination of nature by the understanding in accordance with this unity. This unity could only ever be justified as a transcendental condition of scientific empirical enquiry as relating to an undetermined idea of systematicity of nature.

Kant describes the structure of laws as 'all alike' insofar as they impose an intelligible structure. Kant establishes a strong analogy between laws of nature and freedom in the *Groundwork*: '*act as if the maxim of your action were to become by your will a* **universal law of nature**'.[80] In this sense, the capacity to deploy laws in the context of nature as regulative ideals, which establish unity for scientific enquiry, does not threaten Kant's account of morality. Instead, it offers support to morality insofar as the rational capacity to view nature in accordance with lawlike regularity is made possible by the same faculty of reason that can generate universal maxims for our actions (i.e., categorical imperatives). Much like the relationship between physical and moral teleology,[81] the regulative use of reason for scientific enquiry offers corroboration, but not confirmation, for our moral deployment of reason.

Kant's account of morality raises a problem for Dupré's conception of agency as Dupré does not sufficiently account for the ability for humanity to set itself goals that can be achieved at the societal level. The activities that Dupré identifies are not examples of moral actions under Kant's account. The issue is that such activities could be motivated either by a sense of duty or for some other self-promoting interest. For instance, consider Kant's shopkeeper who finds it advantageous to serve customers honestly.[82] Kant argues that it is not possible to know whether the shopkeeper is acting in accordance with duty or acting in accordance with their own self-interests towards increasing future financial security.

On Dupré's account, it seems either of these motives are sufficient for his account of freedom. One crucial feature of Dupré's account of freedom is that some actions are not reducible to biologically reductive explanations. Importantly, Kant is not claiming it is possible to distinguish between actions that originate from freedom and those that do not in every case. Rather, his aim is to show that we cannot disprove the possibility of freedom. According to Wood, 'Kant does not pretend to know how free agency is possible, but claims only to show that the impossibility of freedom is forever indemonstrable'.[83] Kant's moral philosophy allows us to distinguish between actions arising from duty and those arising from self-interest, even though in practice it we cannot know for certain that actions are genuinely motivated by duty. Dupré's general focus is on changes that have been beneficial for human societies, such as readily available access to food, healthcare, and education. They offer support to the idea that culture can be the source of major evolutionary change.[84]

The problem that is not considered by Dupré, but is essential for Kant's political philosophy, is explaining why we *ought* to act in a way to bring about beneficial societal changes. For instance, Kant argues that for every individual to receive the benefits of healthcare and education, we must first presuppose a unified will of the people to develop a society that will perpetually maintain itself. In such a society, it is the responsibility of the state to 'maintain those members of society who are unable to maintain themselves'.[85] Kant suggests that the state ought to achieve this by imposing constraints on the wealthy individuals of society to provide basic support for those in need. The wealthy are obliged to fulfil this role for the state, 'since they owe their existence to an act of submitting to its protection and care, which they need in order to live'.[86]

Dupré does not explicitly advocate that these forms of societal change ought to be universal. However, he does argue that they are benefits that the majority of us receive. The similarity between Kant and Dupré is that both are advocating the importance of human activity in a way that cannot be accounted for under the conditions of biological determinism. Kant's account diverges from Dupré's insofar as it is concerned with how a society ought to develop in a way that preserves the freedoms of its citizens.[87] Kant's explanation of how a society ought to develop draws from the idea that society should strive to achieve the same values as his moral philosophy. According to Kant:

> By the well-being of a state is understood ... that condition in which its constitution conforms most fully to principles of right; it is that

condition which reason, *by a categorical imperative*, makes it obligatory for us to strive after.[88]

This difference between Kant and Dupré is further elucidated by appealing to Kant's distinction between moral anthropology and the metaphysics of morals. The former examines the subjective context that can either hinder or aid societies towards the fulfilment of laws that constitute a metaphysics of morals. Anthropology cannot be the source of a metaphysics of morals for society, although Kant suggests that a metaphysics of morals can be subsequently applied to anthropology.[89] A metaphysics of morals must be separated from anthropology because its laws cannot be derived from nature; rather, it must be based on the moral sense of duty possessed by every citizen.[90] Kant's political philosophy demands the separation of nature and freedom that constitutes the division between theoretical and practical philosophy. Dupré's account is closer to an anthropological account as he focuses on how society has developed in a way that promotes the well-being of its citizens. He does not consider the additional normative claim that society and science *ought* to be engaging in practices directed towards these goals.

On a related matter, Mackie's argument from queerness could be applied to Kant's political philosophy. Recall that Mackie argued against Kant's moral philosophy on the basis that something cannot be both prescriptive and objective.[91] Objective statements about what is the case can be established only in relation to empirical natural facts. As these aspects of Kant's political philosophy are intended to be derived from the principles of Kant's moral philosophy, Mackie's argument could also be extended to Kant's political philosophy. Christine Korsgaard summarises and rejects Mackie's argument against objective values as follows:

> Knowledge of them, Mackie said, would have to provide the knower with both a direction and a motive ... Of course there are entities that meet these criteria. It's true that they are queer sorts of entities, and that knowing them isn't like anything else. But that doesn't mean that they don't exist.[92]

In the context of Kant's political philosophy, the intuitive appeal of Korsgaard's treatment of Mackie's argument is compelling. If the policies that societies establish are developed in accordance with the moral sense of duty, then they are both normative and objective. The core of Mackie's argument is essentially that facts can be objective, but values cannot. Dupré argues that values are evaluative claims that hold some interest for parts of society, whereas facts possess little interest for society: 'if most or all of physics is value-free, it is not because physics is science, but because most of physics simply doesn't matter to us.'[93] Dupré does not intent to purge values from science, but rather to understand 'how normativity finds its way into scientific work, and how its denial can be potentially dangerous'.[94]

It is important to make explicit both the values that can motivate scientific research and the implicit values that can affect the methodology of research. For instance, Dupré considers how research into the evolution of rape has appealed to similar behaviours in nature exhibited by flies or ducks. He argues that the lack of biological similarities between these entities reveals how such connection is insufficient; 'only in the crudest analogical sense could ... the behaviour of copulating flies be related to that of human rapists'.[95] Appeals to this kind of argument have become less frequent in contemporary evolutionary theory, but the underlying point is that such examples emerge because researchers do not adequately identify the relevant factors that dispose individuals to these behaviours in different species. For humans, rape is inherently a normative concept in both the way it is defined and the motivations for these actions. According to Dupré:

> Those who have thought seriously about contemporary sexual violence as opposed to hypothetical reproductive strategies of imagined ancestors have observed that rape ... has more to do with misogyny, and more to do with violence than sex, let alone reproduction. Its causes appear therefore to be at the level of ideology rather than economics.[96]

Dupré demonstrates both how values are inseparable from science and the need to scrutinise the inherent normative motivation of scientific methodology. Science is not a domain of factual statements that do not matter for us. The importance of values in science has implications for naturalism. The sociobiological project that Dupré is critiquing attempted to provide biological explanations for the emergence of behaviours. Despite its misguided approach to reducing all behaviours to biological explanations, Dupré commends the sociobiological account because of its emphasis on the distinction between something being natural and it being good. Even if it were possible to provide a naturalistic explanation of rape, this clearly would not entail that by extension rape is good. In contrast, it is good that research aimed at discovering the factors that dispose individuals towards rape can, if successful, be used to inform policy aimed at prevention. The problem is that these factors are difficult to identify, especially if researchers do not consider the relevant normative contexts that dispose individuals towards these behaviours. It should be noncontroversial to argue that significant portions of science are conducted because they reflect broader societal interests, which could be directed toward promoting societal safety, securing future commercial opportunities, etc. Moreover, in many cases an accurate understanding of the relevant factors specific to current scientific knowledge will require an examination of the values that have framed, and continue to frame, our scientific understanding.

This account is largely consistent with Kant's political philosophy. The well-being of a society would require the state to be continually vigilant against the abuse of the rights of its citizens. In fact, Kant's account provides justification for Dupré's implicit claim that the development of science should be developing towards socially progressive aims. The relation that Kant establishes between his moral and political philosophy allows a Kant-inspired approach to explain why research *ought* to be directed towards influencing policies that increase the protection of the state against potential harm. In the context of Kant's critical philosophy, this would require us to regard the objects of our scientific understanding as guided by the regulative ideas of reason, not as objects in themselves. This is necessary to account for how humans can develop moral maxims in accordance with freedom from an atemporal and non-natural standpoint. This is not because of the determinacy of nature, but rather because the reason why actions are deemed of moral worth is because they are generated by the capacity of humans to set themselves ends in accordance with duty at both the individual and societal levels.

Conclusion

Kant's philosophy can potentially support and offer guidance to aspects of contemporary philosophy of biology. Kant offers a critical analysis of the assumption that the biological sciences are thoroughly grounded on the principles of naturalism alone. Conditions of biological individuality such as genetic homogeneity and functional integration offer alternative accounts of the boundaries and number of biological individuals that are, at times, incompatible. The lack of consensus regarding a single definition of biological individuality exposes the importance of judgement within biology, and exposes the difficulties for relationship between biology and naturalism.

Appeals to Kant's philosophy by contemporary philosophers of biology reveal the complex relationship between biology and naturalism. The fundamental difference between Kant's account of teleological judgement and contemporary accounts of biological autonomy is that Kant regarded physical teleology as a product of judgement, which is ultimately dependent on our capacity for moral teleology, whereas the contemporary account of biological autonomy does not regard organisms as the product of any form of judgement. In the context of Kant's critical philosophy, the capacity to view nature in accordance with physical teleology is dependent on our capacity for moral teleology. This aspect of Kant's philosophy remains largely underemphasised in Kant scholarship, which contributes to Kant's appeal for these philosophers of biology. However, it is crucial to understand the intricate relationship between moral and physical teleology within Kant's philosophy. That it is possible to conceive of nature teleologically does not offer additional support to the conclusions of the moral proof. However, it offers corroboration for moral teleology. Kant thought that natural teleology had a close relationship with both moral teleology and physico-theology. It follows that these appeals to Kant in support of biology are problematic due the tension between their commitment to naturalism and Kant's non-naturalistic justification of judgements of physical teleology.

In the final section of this chapter, I explored potential ways that Kant's philosophy might offer support and guidance to the future direction of scientific research. I focused on Dupré's appeal to Kant in support of his conception of biological freedom. Dupré argued that Kant's account was along the right lines because he showed that the organism was the source of its own agency. Kant offers an alternative solution to the tendency in philosophy of biology to reduce explanations of morality to emergent properties of evolution. A Kant-inspired approach can offer support to aspects of Dupré's account and highlight the benefits of including societal-level developments and individual human agency within biological explanations. For Dupré, biological accounts that focus on genetic reductivism will be unable to explain how societal developments can be sources of evolutionary change. He considers access to healthcare and education as examples of this. Dupré's account focuses implicitly on societal developments that are supported by aspects of Kant's account of political philosophy. For instance, Dupré suggested that it is good that science develops in a way that promotes social progression and discourages citizens from engaging in behaviours that are detrimental for its citizens, such as rape. This is compatible with Kant's political philosophy, which argues for the state's responsibility to ensure a level of protection for its citizens and support those members of society that cannot support themselves. Kant's political philosophy offers the foundations for the view that society *ought* to develop in these ways.

If some philosophers of science desire a theory that not only recognises that values are inherently related to scientific research, but also makes it possible to direct scientific research towards the promotion of socially progressive values, then Kant's political philosophy offers one way of showing how this can be achieved. Philosophers of science need to relinquish their strong commitment to naturalism in order to recognise the contribution of values for scientific theories and to provide a normative framework that can encourage scientific research towards these values.

Conclusion

Kant's philosophy has been a significant source of influence for both the development of biology and contemporary philosophers of biology. The incompatibilities between biological naturalism and transcendental idealism reveal that in many cases, Kant's influence on these scientific developments cannot be sufficiently justified from the perspective of biological naturalism. I have demonstrated how Kant's philosophy was influential in many cases, including: Whewell's conception of consilience; the design of nature understood as a heuristic principle; and contemporary conceptions of biological autonomy.

In my discussion of these cases, I have revealed how philosophers of biology have repeatedly appealed to Kant to resolve issues that go beyond their commitments to biological naturalism. There has also been significant interest in developing interpretations of Kant that offer support to contemporary philosophy of science. This is strong evidence for the importance of the Kantian perspective for contemporary issues in philosophy of science. However, I have argued that Kant's philosophy provides important critiques, as well as support, for contemporary issues in philosophy of biology. One motivation for writing this book was that I accepted Zammito's argument regarding the incompatibility between transcendental idealism and biological naturalism, yet I was unconvinced by his conclusion that Kant was not significantly influential on biology. I proposed a new approach to influence that distinguished between an influence occurring and the requirement to correctly understand the source of the influence. I saw potential insight for biology more generally by examining precisely how Kantian principles had informed biology, and the broader implications of this. I applied this approach in my discussion of the importance of Hume's philosophy for Kant in Chapter 2, and the significance of Kant's philosophy for Whewell in Chapter 3. In both cases, I argued that the emerging theories were not consistent with the original theories. However, by examining the role that the original theories played for the emerging theories, I developed a new perspective that identified some motivating factors behind these accounts.

I applied my interpretation of Kant's philosophy to contemporary debates in philosophy of science throughout this book. I constructed recent debates regarding the metaphysical status of the laws of nature into a Kantian mathematical antinomy at the end of Chapter 2. In Chapter 4 I argued that recent explanations of biological design appeal, either explicitly or implicitly, to a regulative account of the organism as a product of judgement like the one developed by Kant. Kant's philosophy of science has a great deal to offer contemporary philosophy of science, but this would require a re-evaluation of some of commonly held principles in philosophy of science. A Kant-inspired philosophy of science could only offer justification for an account of science that understood scientific development in accordance with the regulative demands of reason. The highest demand of reason would be the need to presuppose the regulative principle of unity as the motivation for science, without requiring that this unity related to any object independent of experience. It would require the separation of the conditions of knowledge of appearances from objects in themselves, and for scientists and philosophers of science to accept that our scientific knowledge always relates to appearance.

I describe this as a Kant-inspired philosophy of science, rather than a Kantian philosophy of science, because Kant regarded proper science as apodictically certain and universal. However, scientific developments have demonstrated that the level of certainty and universalisability achieved by Newtonian science, which Kant considered as a paradigm of proper science, is an exceptional case that is not representative of science more generally. Cartwright opposes the view that scientific laws are universal and apply across nature, instead she argues that the regularity achieved under experimental conditions cannot be generalised to nature as a whole. She describes the creation of these regularities as nomological machines, which draws attention to how these context-specific regularities depend on scientists to create the experimental conditions. Potochnik also emphasised the how factors outside of science can be significant for scientific knowledge. She argued that our scientific knowledge is limited due to our biological conditions such as limited lifespan, and from social and political factors that influence the development of science. Our scientific knowledge has been due to our ability to apply idealisations to nature to isolate specific causal patterns. Historically these idealisations have been associated with discovering knowledge about the metaphysical laws of nature, but Potochnik argues they reflect our biological limitations and sociopolitical context of the development of science.

Both the continued significance of Kant's philosophy for biology and his influence on the development of biology have created the opportunity to examine the role of naturalism within biology. I have argued that certain developments in philosophy of biology do not fit well with some of the general assumptions behind biological naturalism. The growing support within philosophy of biology towards accounts that recognise the significance of social and political factors are indicative of this tension.

The dichotomy between naturalism and transcendental idealism is often raised by those who believe that we can explain broader aspects of humanity from the standpoint of naturalism alone. Wilson, Mackie and Hull all demonstrated their strong commitment to this account of biological naturalism, as they believed that we would discover the biological foundation to explain the emergence of morality. If naturalism is understood as an exhaustive reductive explanation of nature, then there can be no reconciliation between naturalism and Kantian transcendental idealism. Ironically, Wilson drew from the principle of consilience as justification for this belief, which I have argued emerged in part due to Kant's influence on Whewell. However, the alternative human-centric account of science seems more closely aligned to elements of Kant's philosophy. From this perspective, there is little hope for a reductive account of biology to uncover an exhaustive explanation of our behaviours and moral values. On the contrary, I have argued that values are an irreducible element of the milieu that continues to shape biology.

I have explored how we might derive some general governing principles for biology from Kant's political philosophy. The architectonic structure of Kant's critical philosophy provides some indication of how politics and science might converge to develop shared principles. I argued that Dupré implicitly appealed to the social and political responsibilities of science in his critical evaluation of biological explanations of rape in humans. The complexity of factors that motivate human behaviours means that it is deeply misguided to consider explanations of similar behaviours in flies and ducks. The danger is that if our scientific explanations do not correctly identify the factors motivating these actions, then this research will not effectively inform policies to potentially prevent these actions in the future and offer protection for citizens. Elements of Kant's political philosophy discuss social responsibilities of the state to ensure protection for all citizens, but these discussions are not sufficiently developed within his *Metaphysics of Morals*. I suggest we draw inspiration from Kant's architectonic approach to his critical philosophy as a whole, which investigated how various disciplines could come together including: epistemology, science, metaphysics, religion, ethics, politics, etc. It is important for philosophers of science who assert the importance of social and political factors for the development of science to consider the extent to which it is possible to foster productive relationships and identify shared objectives between science and politics. By no means does Kant offer a complete answer to this question, but I hope to have laid the foundations for future discussions about how to understand biology within its broader social and political contexts.

Notes

Introduction

1 Immanuel Kant, *Critique of Pure Reason*, Paul Guyer and Allen W. Wood (trans.) (Cambridge: Cambridge University Press, 1998), B73.

2 Immanuel Kant, *Metaphysical Foundations of Natural Science*, Michael Friedman (ed. and trans.) (Cambridge: Cambridge University Press, 2004), 4: 468.

3 William Coleman, *Biology in the Nineteenth Century: Problems of Form, Function and Transformation* (Cambridge: Cambridge University Press, 1977), p. 1.

4 John Zammito, 'Teleology Then and Now: The Question of Kant's Relevance for Contemporary Controversies over Function in Biology', *Studies in History and Philosophy of Science Part C: Studies in History and Philosophy of Biological and Biomedical Sciences*, 37/4 (2006), 748–70.

5 Scott Lidgard and Lynn K. Nyhart, 'The Work of Biological Individuality: Concepts and Contexts', in Scott Lidgard and Lynn K. Nyhart (eds), *Biological Individuality: Integrating Scientific, Philosophical, and Historical Perspectives* (Chicago IL: University of Chicago Press, 2017), p. 27.

6 Kant, *Critique of Pure Reason*, A84/B116.

7 Zammito, 'Teleology Then and Now', 766.

8 Richard Lewontin, *The Triple Helix: Gene, Organism and Environment* (Cambridge MA: Harvard University Press, 2000), p. 3.

9 Lewontin, *The Triple Helix*, p. 4.

10 Kant, *Critique of Pure Reason*, Axi–Axii.

11 Stephen Engstrom, 'Knowledge and Its Object', in James R. O'Shea (ed.), *Kant's Critique of Pure Reason: A Critical Guide* (Cambridge: Cambridge University Press, 2017), p. 28.

Chapter 1

1 I understand the biological commitment to naturalism broadly as upholding the view that the domain of the biological sciences can be

explained exclusively by natural means. 'Naturalism' is notoriously difficult to define; for instance, consider Barry Stroud's comparison between naturalism and world peace: 'Almost everyone swears allegiance to it, and is willing to march under its banner. But disputes can still break out about what it is appropriate or acceptable to do in the name of that slogan. And like world peace, once you start specifying concretely exactly what it involves and how to achieve it, it becomes increasingly difficult to reach and to sustain a consistent and exclusive "naturalism."' Barry Stroud, 'The Charm of Naturalism', *Proceedings and Addresses of the American Philosophical Association*, 70/2 (1996), 43–4.

2 John Zammito, 'The Lenoir Thesis Revisited: Blumenbach and Kant', *Studies in History and Philosophy of Science Part C: Studies in History and Philosophy of Biological and Biomedical Sciences*, 43/1 (2012), 120–32.

3 Zammito, 'Teleology Then and Now', 749.

4 Kant, *Critique of Pure Reason*, Axxi.

5 Peter Strawson, *The Bounds of Sense: An Essay on Kant's Critique of Pure Reason* (London: Methuen and Co., 1966), pp. 40–1.

6 Strawson, *The Bounds of Sense*, p. 252.

7 Kant, *Critique of Pure Reason*, A30/B45.

8 Kant, *Critique of Pure Reason*, A36/B52.

9 Strawson, *The Bounds of Sense*, p. 242.

10 Lucy Allais, *Manifest Reality: Kant's Idealism and His Realism* (Oxford: Oxford University Press, 2015), p. 9.

11 Allais, *Manifest Reality*, p. 219.

12 John Dupré, *Processes of Life: Essays in the Philosophy of Biology* (Oxford: Oxford University Press 2012), p. 30.

13 John Dupré and Daniel J. Nicholson, 'A Manifesto for a Processual Philosophy of Biology', in *Everything Flows: Towards a Processual Philosophy of Biology* (Oxford: Oxford University Press, 2018), p. 4.

14 Immanuel Kant, 'The Metaphysics of Morals', in Mary J. Gregor (ed.), *Practical Philosophy* (Cambridge: Cambridge University Press, 1996), 6: 375.

15 Strawson, *The Bounds of Sense*, p. 248.

16 This is from Norman Kemp Smith's translation as the term '*Erkenntnis*' is translated as knowledge, whereas Paul Guyer and Allen Wood translate this as cognition.

17 Kant, *Critique of Pure Reason*, B158.

18 The difference between the transcendental and biological accounts of epigenesis is emphasised by Marcel Quarfood: '[n]obody believes that transcendental idealism is based on geography just because CPR contains geographical analogies, and likewise it should be clear that

an epigenetical account of the categories is not part of the biological theory of epigenesis.' Marcel Quarfood, *Transcendental Idealism and the Organism: Essays on Kant* (Stockholm: Almqvist & Wiksell International, 2004), p. 102.

19 Kant, *Critique of Pure Reason*, B167.

20 Jennifer Mensch, *Kant's Organicism: Epigenesis and the Development of Critical Philosophy* (Chicago IL: University Of Chicago Press, 2013), p. 139.

21 Kant, *Critique of Pure Reason*, A86-7/B119.

22 Stuart A. Kauffman, *Investigations* (Oxford: Oxford University Press, 2000), p. 50.

23 Immanuel Kant, *Critique of the Power of Judgment*, Paul Guyer (ed.), Paul Guyer and Eric Matthews (trans.) (Cambridge: Cambridge University Press, 2000), 5: 400.

24 Mensch, *Kant's Organicism*, p. 144.

25 Immanuel Kant, 'Review of J. G. Herder's Ideas for the Philosophy of the History of Humanity. Parts 1 and 2 (1785)', in Günter Zöller and Robert B. Louden (eds), *Anthropology, History, and Education*, Allen W. Wood (trans.) (Cambridge: Cambridge University Press, 2007), 8: 56.

26 Thomas Kuhn, *The Structure of Scientific Revolutions*, Enlarged edition (Chicago IL: Chicago University Press, 1970), p. 204.

27 Henry E. Allison, 'Transcendental Idealism and Descriptive Metaphysics', *Kant-Studien*, 60/2 (1969), 224.

28 Allison, 'Transcendental Idealism and Descriptive Metaphysics', 231.

29 Kant, *Critique of Pure Reason*, A491/B519.

30 Kant, *Critique of Pure Reason*, A749/B777.

31 Kant, *Critique of Pure Reason*, A750/B778.

32 Allison, 'Transcendental Idealism and Descriptive Metaphysics', 217.

33 Henry E. Allison, *Custom and Reason in Hume: A Kantian Reading of the First Book of the Treatise, Custom and Reason in Hume* (Oxford: Oxford University Press, 2008), p. 279.

34 Allison, *Custom and Reason in Hume*, p. 279.

35 Immanuel Kant, 'What Does It Mean to Orient Oneself in Thinking?', in Allen W. Wood and George di Giovanni (eds), *Religion and Rational Theology* (Cambridge: Cambridge University Press, 1996), 8: 136.

36 Kant does engage with metaphysical questions regarding the freedom, immortality and the soul; however, these transcendental enquiries do not require that we consider these entities as appearances. Instead, they are ideals of reason.

37 Strawson, *The Bounds of Sense*, p. 272.

38 Kuhn's account is of additional importance for this investigation not only because of his contribution to this debate, but also because

Lenoir's account of Kant's influence on the development of German biology in the nineteenth century draws from an account of scientific methodology that is indebted to Kuhn (see section 1.3).

39 Isaiah Berlin, *The Proper Study Of Mankind: An Anthology of Essays* (London: Pimlico, 1998), p. 27.

40 Berlin, *The Proper Study Of Mankind*, p. 27.

41 Berlin, *The Proper Study Of Mankind*, p. 27.

42 Berlin, *The Proper Study Of Mankind*, p. 57.

43 Kuhn, *The Structure of Scientific Revolutions*, p. 24.

44 Thomas S. Kuhn, 'Logic of Discovery or Psychology of Research?', in Imre Lakatos and Alan Musgrave (eds), *Criticism and the Growth of Knowledge: Proceedings of the International Colloquium in the Philosophy of Science* (Cambridge: Cambridge University Press, 1970), p. 6.

45 Paul K. Feyerabend, 'Consolations for the Specialist', in Imre Lakatos and Alan Musgrave (eds), *Criticism and the Growth of Knowledge: Proceedings of the International Colloquium in the Philosophy of Science* (Cambridge: Cambridge University Press, 1970), p. 200.

46 Berlin, *The Proper Study Of Mankind*, p. 31.

47 Berlin, *The Proper Study Of Mankind*, p. 32.

48 Kuhn, 'Logic of Discovery or Psychology of Research?', p. 137.

49 Kuhn, *The Structure of Scientific Revolutions*, p. 140.

50 Ernst Mayr, *The Growth of Biological Thought: Diversity, Evolution and Inheritance* (Cambridge MA: Harvard University Press, 1982), p. 56.

51 Michael Ruse, 'Reply to Richards', in *Debating Darwin* (Chicago IL: University of Chicago Press, 2016), p. 181.

52 Robert J. Richards, 'Charles Darwin: Cosmopolitan Thinker', in *Debating Darwin* (Chicago IL: University of Chicago Press, 2016), p. 115.

53 Strawson, *The Bounds of Sense*, p. 11.

54 Robin G. Collingwood, *An Essay On Metaphysics* (Oxford: Clarendon Press, 1940), p. 57.

55 This raises an important problem for some of the claims proposed in this work that suggest Kant's philosophy holds important insight for contemporary issues in philosophy of biology. It is helpful to distinguish between strong and weak versions of Collingwood's argument from historicity. The strong argument outlined above denies any possibility for the discovery of universal principles. The weak argument allows for the possibility of the discovery of certain universal principles (say, for instance, the conditions for the possibility of experience in general), which is in contrast to the historical embeddedness of scientific theories. The simplest argument for endorsing the weak argument

for historicity when examining Kant's influence on the development of biology is that philosophers of biologists continue to appeal to Kant's philosophy to elucidate historical and contemporary issues.

56 Kuhn, *The Structure of Scientific Revolutions*, p. 84.

57 Kuhn, *The Structure of Scientific Revolutions*, p. 153.

58 Kuhn, *The Structure of Scientific Revolutions*, p. 98.

59 Feyerabend, 'Consolations for the Specialist', p. 202.

60 Gottfried Wilhelm Leibniz, *Philosophical Essays*, Roger Ariew and Daniel Garber (eds) (Indianapolis IN: Hackett, 1989), p. 214.

61 Kuhn, *The Structure of Scientific Revolutions*, p. 193. Kuhn's claims regarding the untranslatability of scientific paradigms, and that members of different paradigms live in different worlds, have been criticised by Donald Davidson. He argues that objective truth must relate to a world that is partially translatable between different languages and other kinds of conceptual scheme. According to Davidson, translation can fail only if 'there is something neutral and common that lies outside all schemes'. Donald Davidson, 'On the Very Idea of a Conceptual Scheme', *Proceedings and Addresses of the American Philosophical Association*, 47 (1973), 12.

62 Kant, *Critique of Pure Reason*, Bxxii.

63 Michael Friedman, 'Kant, Kuhn, And The Rationality Of Science', *Philosophy of Science*, 69/2 (2002), 183.

64 Kuhn, *The Structure of Scientific Revolutions*, p. 92.

65 What is meant by the 'evolution of political institutions' is not clear.

66 Reidar Maliks, *Kant's Politics in Context* (Oxford: Oxford University Press, 2014), p. 122.

67 Kant, 'The Metaphysics of Morals', 6: 355.

68 Feyerabend, 'Consolations for the Specialist', p. 198.

69 Kuhn, *The Structure of Scientific Revolutions*, p. 151.

70 Kuhn, *The Structure of Scientific Revolutions*, p. 82.

71 Kuhn, *The Structure of Scientific Revolutions*, p. 64.

72 Kuhn, *The Structure of Scientific Revolutions*, p. 96.

73 Imre Lakatos, 'Falsification and the Methodology of Scientific Research Programmes', in Alan Musgrave and Imre Lakatos (eds), *Criticism and the Growth of Knowledge: Proceedings of the International Colloquium in the Philosophy of Science* (Cambridge: Cambridge University Press, 1970), p. 135.

74 Lakatos, 'Falsification and the Methodology of Scientific Research Programmes', p. 175.

75 Timothy Lenoir, *The Strategy of Life: Teleology and Mechanics in Nineteenth-Century German Biology* (Dordrecht: Reidel Publishing Company, 1982), p. 13.

76 Lenoir, *The Strategy of Life*, p. 278.

77 Lynn K. Nyhart, *Biology Takes Form: Animal Morphology and the German Universities, 1800–1900* (Chicago IL: University of Chicago Press, 1995), p. 8.
78 Lakatos, 'Falsification and the Methodology of Scientific Research Programmes', p. 142.
79 Lakatos, 'Falsification and the Methodology of Scientific Research Programmes', p. 143.
80 Immanuel Kant, *Correspondence*, Arnulf Zweig (ed.) (Cambridge: Cambridge University Press, 1999), 11: 185.
81 Kant, *Critique of the Power of Judgment*, 5: 424.
82 Robert J. Richards, *The Romantic Conception of Life: Science and Philosophy in the Age of Goethe* (Chicago IL: University of Chicago Press, 2002), p. 229.
83 Lenoir, *The Strategy of Life*, p. 53.
84 Lenoir, *The Strategy of Life*, p. 2.
85 Zammito, 'Teleology Then and Now', 749.
86 Harold Bloom, *The Anxiety of Influence: A Theory of Poetry*, 2nd edition (New York: Oxford University Press, 1997), p. 26.
87 Bloom, *The Anxiety of Influence*, p. 30.
88 Bloom, *The Anxiety of Influence*, p. 44.
89 P. K. Feyerabend, *Against Method* (London: Verso, 1978), p. 165
90 Hans Reichenbach, *Experience and Prediction: An Analysis of the Foundations and the Structure of Knowledge* (Chicago IL: The University of Chicago Press, 1938), p. 384.
91 Karl R. Popper, *The Logic of Scientific Discovery* (London: Routledge, 2002), p. 7.
92 Reichenbach, *Experience and Prediction*, p. 9.
93 Angela Potochnik, *Idealization and the Aims of Science* (Chicago IL: University of Chicago Press, 2017), p. 94.
94 Potochnik, *Idealization and the Aims of Science*, p. 19.
95 Potochnik, *Idealization and the Aims of Science*, p. 94.
96 Potochnik, *Idealization and the Aims of Science*, p. 94.
97 Potochnik, *Idealization and the Aims of Science*, p. 203.
98 This point is important for the discussion of the unity of science discussed in Chapter 3 (section 3.2).
99 Collingwood, *An Essay On Metaphysics*, p. 68.
100 Collingwood, *An Essay On Metaphysics*, pp. 193–4.
101 Collingwood, *An Essay On Metaphysics*, p. 194.
102 See Chapter 1 (section 1.2.2).
103 Although he claims that Greek philosophers from Thales were almost entirely monotheistic.
104 Collingwood, *An Essay On Metaphysics*, p. 207.
105 Collingwood, *An Essay On Metaphysics*, pp. 206–7.

106 God's role in the development of science will be returned to in Chapter 3 when discussing Kant's influence on the development of biology in Britain.

107 Feyerabend, *Against Method*, pp. 153–4.

108 This perspective also has no difficulty understanding how Kant was inspired by the biological account of epigenesis to develop the position that the emergence of reason was epigenetic. This was not compatible with the biological account of epigenesis, yet it was still influential on the development of transcendental idealism. This was discussed in Chapter 1 (section 1.1.2).

109 Zammito, 'Teleology Then and Now', 766.

110 Zammito, 'Teleology Then and Now', 765. Analogously, on the relation between Kant and Blumenbach, Zammito asserts 'Blumenbach's affiliation with Kant is best understood as a misunderstanding. But it was a creative misunderstanding, because it enabled Blumenbach and his followers to continue with even greater energy the development of that new science'. Zammito, 'The Lenoir Thesis Revisited', 127. Similarly he claims 'Blumenbach's affiliation with Kant is best understood as a *misunderstanding* – though an influential one in the constitution of biology as a discipline in the succeeding decades'. John H. Zammito, *The Gestation of German Biology: Philosophy and Physiology from Stahl to Schelling* (Chicago IL: University of Chicago Press, 2018), p. 236.

Chapter 2

1 This term can also be translated as memory, recollection, reminder or reminiscence.

2 Immanuel Kant, *Prolegomena to Any Future Metaphysics: With Selections from the Critique of Pure Reason*, Gary Hatfield (eds) (Cambridge: Cambridge University Press, 1997), 4: 260.

3 Kant, *Prolegomena to Any Future Metaphysics*, 4: 260.

4 Karin de Boer, 'Kant's Response to Hume's Critique of Pure Reason', *Archiv für Geschichte der Philosophie*, 101/3 (2019), fn. 1.

5 Manfred Kuehn, *Kant: A Biography* (Cambridge: Cambridge University Press, 2001), p. 185.

6 Reed Winegar, 'Kant's Criticisms of Hume's Dialogues Concerning Natural Religion', *British Journal for the History of Philosophy*, 23/5 (2015), 889.

7 Beattie's *Essays on the Nature and Immutability of Truth* was translated into German in 1772.

8 Lewis White Beck, 'A Prussian Hume and a Scottish Kant', in *Essays on Kant and Hume* (London: Yale University Press, 1978), p. 120.

9 Beck, 'A Prussian Hume and a Scottish Kant', p. 120.

10 Paul Guyer, *Knowledge, Reason, and Taste: Kant's Response to Hume* (Princeton NJ: Princeton University Press, 2008), p. 76fn.

11 Manfred Kuehn, 'Kant's Conception of "Hume's Problem"', *Journal of the History of Philosophy*, 21/2 (1983), 189.

12 To be clear, this does not mean that Beattie's account had no influence on Kant; rather Kant's interpretation was the combined result of Beattie, Hume's *Enquiry* and his limited access to the *Treatise*.

13 Eric Watkins, *Kant and the Metaphysics of Causality* (Cambridge: Cambridge University Press, 2005), p. 386.

14 Guyer, *Knowledge, Reason, and Taste*, p. 20.

15 Allison, *Custom and Reason in Hume*, p. 356fn.1.

16 Kant, *Critique of Pure Reason*, A761/B788.

17 David Hume, *Enquiries Concerning Human Understanding and Concerning the Principles of Morals*, L. A. Selby-Bigge and P. H. Nidditch (eds), Third edition (Oxford: Oxford University Press, 1975), 22.

18 Hume, *Enquiries Concerning Human Understanding and Concerning the Principles of Morals*, 26.

19 Kant, *Critique of Pure Reason*, A84/B116.

20 Hume, *Enquiries Concerning Human Understanding and Concerning the Principles of Morals*, 36.

21 Immanuel Kant, 'What Real Progress Has Metaphysics Made in Germany since the Time of Leibniz and Wolff? (1793/1804)', in Gary Hatfield and Henry Allison (eds), *Theoretical Philosophy after 1781*, Michael Friedman and Peter Heath (trans.) (Cambridge: Cambridge University Press, 2002), 20: 266.

22 Kant, *Prolegomena to Any Future Metaphysics*, 4: 272.

23 Hume, *Enquiries Concerning Human Understanding and Concerning the Principles of Morals*, 131.

24 Immanuel Kant, 'Critique of Practical Reason', in Mary J. Gregor (ed.), *Practical Philosophy* (Cambridge: Cambridge University Press, 1996), 5: 53.

25 Hume, *Enquiries Concerning Human Understanding and Concerning the Principles of Morals*, 118.

26 Hume, *Enquiries Concerning Human Understanding and Concerning the Principles of Morals*, 119.

27 Hume, *Enquiries Concerning Human Understanding and Concerning the Principles of Morals*, 119.

28 Kant, *Critique of Pure Reason*, A375–A376.

29 Kant, *Critique of Pure Reason*, A477/B505 and A763/B791.

30 David Hume, *A Treatise of Human Nature*, L. A. Selby-Bigge and P. H. Nidditch (eds) Second edition (Oxford: Oxford University Press, 1978), 269.

31 Paul Guyer, 'Imperfect Knowledge of Nature', in Angela Breitenbach and Michela Massimi (eds), *Kant and the Laws of Nature* (Cambridge: Cambridge University Press, 2017), p. 67.

32 Kant, *Critique of Pure Reason*, B19.

33 Kant, *Prolegomena to Any Future Metaphysics*, 4: 273.

34 Kant, 'Critique of Practical Reason', 5: 52.

35 Donald Gotterbarn, 'Kant, Hume and Analyticity', *Kant-Studien*, 65/1–4 (1974), 279.

36 The Cambridge translations of both the *Prolegomena* (4: 273) and the first *Critique* (B19) are accompanied by editor's notes that comment on the difference between Hume's account in the *Treatise* and *Enquiry*. They suggest that Kant had not read the *Treatise* when he produced these texts, yet they maintain that Kant's account is consistent with the *Enquiry*.

37 Beck, 'A Prussian Hume and a Scottish Kant', p. 84.

38 Gotterbarn, 'Kant, Hume and Analyticity', 280–1.

39 Hume, *Enquiries Concerning Human Understanding and Concerning the Principles of Morals*, 131.

40 Gotterbarn, 'Kant, Hume and Analyticity', 281.

41 Kant was incorrect to regard himself as the first person to identify the distinction between synthetic and analytic judgements; moreover, his claim that some experiences were dependent on *a priori* foundations was also not completely original. According to Gottenbarn 'Many philosophers before Kant had made synthetic judgements known *a priori*; such judgements were not the invention of Kant'. Gotterbarn, 'Kant, Hume and Analyticity', 281.

42 Galen Strawson, 'David Hume: Objects and Powers', in *Real Materialism: And Other Essays* (Oxford: Oxford University Press, 2008), p. 418.

43 Kenneth Richman, 'Introduction', in Rupert Read and Kenneth Richman (eds), *The New Hume Debate: Revised Edition* (London: Routledge, 2007), p. 1.

44 Hume, *Enquiries Concerning Human Understanding and Concerning the Principles of Morals*, 44.

45 Gottfried Wilhelm Leibniz, *Philosophical Essays*, Roger Ariew and Daniel Garber (eds) (Indianapolis IN: Hackett, 1989), p. 223.

46 Juliet Floyd, 'The Fact of Judgment: The Kantian Response to the Humean Condition', in Jeff Malpas (ed.), *From Kant to Davidson: Philosophy and the Idea of the Transcendental* (London: Routledge, 2003), p. 37.

47 Hume, *Enquiries Concerning Human Understanding and Concerning the Principles of Morals*, 45.

48 Guyer, *Knowledge, Reason, and Taste*, p. 89.

49 Guyer, *Knowledge, Reason, and Taste*, p. 93.

50 Kenneth Winkler, 'The New Hume', in Rupert Read and Kenneth Richman (eds), *The New Hume Debate*, Revised edition (London: Routledge, 2007), p. 67.

51 Galen Strawson, *The Secret Connexion: Causation, Realism, and David Hume*, Revised edition (Oxford: Oxford University Press, 2014), p. 228. The difference between 'Causation' and 'causation' is that the former is a metaphysical account, whereas the latter is derived from experience as custom or habit.

52 Strawson, *The Secret Connexion*, p. 228–9.

53 This appeal to evolution is highly anachronistic given that Hume was not aware of our current understanding of evolution.

54 Allison, *Custom and Reason in Hume*, p. 142.

55 Kant, *Critique of Pure Reason*, A760/B788.

56 Kant, *Critique of Pure Reason*, A761/B789.

57 Watkins, *Kant and the Metaphysics of Causality*, p. 383.

58 Mensch, *Kant's Organicism*, p. 153.

59 See Chapter 1 (section 1.1.2).

60 The transformation of this concept from a ground of nature to a ground of reason led Kant to argue that organisms could not be regarded as existing independent of judgement. For Kant, organisms are a product of our ability to judge nature *as if* it were created accordance with final causes.

61 The appeal to epigenesis creates a potential conflict with the analogy of reason as developmental. If reason emerged epigenetically, then it already contained this *a priori* structure. This conflict is resolved if we consider the epigenesis of reason in the context of its role in the development of Kant's philosophy, whereas the lineage of reason is meant historically or anthropologically. Reason itself has not changed; only the orientation towards the understanding of reason has changed. Hence, both explanations are compatible when considered from these different perspectives.

62 The appropriateness of 'intuition' has been disputed, as *Anschauung* means literally 'atlooking' or 'atsight' and is closely related to the family of terms 'foresight' or 'insight'. These connote looking at an object in its immediate presence, whereas intuition connotes a direct, yet unexplainable, transference of information. (For a more detailed discussion, see the editorial discussion in *Mind* (1892), 'What does Anschauung Mean?'.)

63 Kant, *Critique of Pure Reason*, A30/B46.

64 Kant, *Critique of Pure Reason*, A31/B46.

65 Allison, *Custom and Reason in Hume*, p. 110.

66 Hume, *A Treatise of Human Nature*, 36.

67 Allison, *Custom and Reason in Hume*, p. 111.

68 Allison, *Custom and Reason in Hume*, p. 54.

69 Paul Guyer, *Kant* (New York: Routledge, 2006), p. 66.

70 Kant, *Critique of Pure Reason*, B146.

71 Allen Wood, Paul Guyer and Henry E. Allison, 'Debating Allison on Transcendental Idealism', *Kantian Review*, 12/2 (2007), 13.

72 Henry E. Allison, *Idealism and Freedom: Essays on Kant's Theoretical and Practical Philosophy* (Cambridge: Cambridge University Press, 1996), p. 25.

73 Guyer, *Kant*, p. 69.

74 This is discussed in more detail in Chapter 5 (section 5.2.3).

75 Immanuel Kant, *Groundwork of the Metaphysics of Morals*, Mary Gregor (ed.), Christine M. Korsgaard (trans.) (Cambridge: Cambridge University Press, 1998), 4: 457.

76 Allen W. Wood, 'Kant's Compatibilism', in Allen W. Wood (ed.), *Self and Nature in Kant's Philosophy*, (London: Cornell University Press, 1984), p. 74.

77 Guyer, *Kant*, p. 68.

78 Henry E. Allison, *Kant's Theory of Freedom* (Cambridge: Cambridge University Press, 1990), p. 40.

79 Allison, *Kant's Theory of Freedom*, p. 40.

80 Kant, 'Critique of Practical Reason', 5: 56.

81 Kant, 'Critique of Practical Reason', 5: 57.

82 Nancy Cartwright, *The Dappled World: A Study of the Boundaries of Science* (Cambridge: Cambridge University Press, 1999), p. 23.

83 This is evident from the title of Kant's *Prolegomena to any Future Metaphysics*.

84 In this sense, Kant is caught in the middle of Bhaskar and Cartwright; he is too empiricist for Bhaskar, but not empiricist enough for Cartwright.

85 Roy Bhaskar, *A Realist Theory of Science* (London: Verso, 2008), p. 40.

86 This is an example of the conflation of Hume's epistemology with his ontology that Galen Strawson opposed. While Hume denied any rational basis for such laws, habit supports our belief in these laws. Hume's rational doubt of universal laws was restricted to a mere academic exercise. Kant's critical philosophy is an attempt to make Hume's philosophy into a substantive theory about the limitations of our human capacities.

87 Bhaskar, *A Realist Theory of Science*, p. 28.

88 Kant, *Critique of Pure Reason*, A719/B747.

89 This criticism was most famously raised by Garve's review of the *Critique of Pure Reason*. See the Cambridge University Press translation of Kant's *Prolegomena to Any Future Metaphysics*, pp. 201–7.

90 Kant, *Prolegomena to Any Future Metaphysics*, 4: 375.

91 Bhaskar, *A Realist Theory of Science*, p. 38.

92 Roy Bhaskar, *Reclaiming Reality: A Critical Introduction to Contemporary Philosophy* (London: Verso, 1989), p. 183.

93 Bhaskar, *A Realist Theory of Science*, p. 36.

94 See Chapter 1 (section 1.1.3).

95 Cf. Kant, *Critique of Pure Reason*, A51/B76

96 Bhaskar, *A Realist Theory of Science*, p. 53.

97 Bhaskar, *A Realist Theory of Science*, p. 78.

98 Bhaskar, *A Realist Theory of Science*, p. 22.

99 Cartwright, *The Dappled World*, p. 31.

100 Cartwright, *The Dappled World*, p. 24.

101 Cartwright, *The Dappled World*, p. 28.

102 John Dupré, *Human Nature and the Limits of Science* (Oxford and New York: Oxford University Press, 2001), p. 166.

103 Cartwright, *The Dappled World*, p. 52.

104 Kant, *Correspondence*, 12: 257–8.

105 James R. O'Shea, *Kant's Critique of Pure Reason: An Introduction and Interpretation* (Durham: Acumen Publishing, 2012), p. 51.

106 Kant, *Critique of Pure Reason*, A530/B558.

107 Kant, *Critique of Pure Reason*, A422-3/B450.

108 Kant, *Critique of Pure Reason*, A424/B451-2.

109 Bhaskar, *Reclaiming Reality*, p. 17.

110 Bhaskar, *A Realist Theory of Science*, p. 52.

111 Kant, *Critique of Pure Reason*, A471/B499.

112 Kant, *Critique of Pure Reason*, A507/B535.

113 Kant, *Critique of Pure Reason*, A503/B531.

114 Kant, *Critique of Pure Reason*, A653/B681.

115 Cartwright, *The Dappled World*, p. 23.

116 Cartwright includes planning, prediction, manipulation, control and policy setting.

117 Stephen Clarke, 'Transcendental Realisms in the Philosophy of Science: On Bhaskar and Cartwright', *Synthese*, 173/3 (2010), 310.

118 Barry Stroud, 'Transcendental Arguments', *The Journal of Philosophy*, 65/9 (1968), 255.

119 Robert Stern, 'Transcendental Arguments: A Plea for Modesty', *Grazer Philosophische Studien*, 74/1 (2007), 145.

120 Stern criticises Stroud for assuming that the difference in modality between claims about existence independent of experience and claims about conditions of experience is sufficient to deny the possibility of transcendental arguments pertaining to the external world without further justification.

Chapter 3

1   Marjorie Grene and David Depew, *The Philosophy of Biology: An Episodic History* (Cambridge: Cambridge University Press, 2004), p. 92.
2   Kant, *Critique of the Power of Judgment*, 5: 195.
3   Kant, *Critique of the Power of Judgment*, 5: 400.
4   Steffen Ducheyne, 'Kant and Whewell on Bridging Principles between Metaphysics and Science', *Kant-Studien*, 102/1 (2011), 35.
5   Robert E. Butts, *Historical Pragmatics: Philosophical Essays* (Dordrecht: Kluwer Academic Publishers, 1993), p. 192.
6   Kant, *Prolegomena to Any Future Metaphysics*, 4: 290.
7   Kant, *Critique of Pure Reason*, A51/B75.
8   Kant, *Critique of Pure Reason*, A245/B303.
9   This was discussed in Chapter 1 (section 1.1.3).
10 William Whewell, *Novum Organon Renovatum*, 3rd edition (London: John W. Parker and Son, 1858), p. 7.
11 Kant, 'Critique of Practical Reason', 5: 53.
12 William Whewell, *The Philosophy of the Inductive Sciences*, 2 vols (London: John W. Parker and Son, 1847), 1, p. 75.
13 Whewell, *Novum Organon Renovatum*, p. 9.
14 Kant, *Critique of Pure Reason*, A28/B44.
15 Eric Watkins, 'Kant on the Distinction between Sensibility and Understanding', in James R. O'Shea (ed.), *Kant's Critique of Pure Reason: A Critical Guide* (Cambridge: Cambridge University Press, 2017), p. 24.
16 Whewell, *The Philosophy of the Inductive Sciences*, 1, p. 18.
17 William Whewell, *On the Philosophy of Discovery* (London: John W. Parker and Son, 1860), p. 489.
18 Kant, *Critique of Pure Reason*, A51/B76.
19 Whewell, *The Philosophy of the Inductive Sciences*, 1, p. 18.
20 The third man argument is originally a criticism of Plato's theory of forms developed in Plato's *Parmenidies* and Aristotle's *Metaphysics*. The argument is that the universal form from which particulars derive their properties must have some common property that is both like the universal and like the particular. Thus, it is necessary to posit a third entity that contains both these properties.
21 For Kant, the role of the schema is to subsume intuitions under concepts. According to Kant, '[t]his mediating representation must be pure (without anything empirical) and yet **intellectual** on the one hand and **sensible** on the other. Such a representation is the **transcendental schema**'. Kant, *Critique of Pure Reason*, A138/B177.
22 Butts, *Historical Pragmatics*, p. 219.
23 W. H. Walsh, 'Schematism', in *Kant: A Collection of Critical Essays*, Robert Paul Wolff (ed.) (New York: Anchor Books, 1967), p. 71.

24 Kant, *Critique of Pure Reason*, A141/B180.
25 Walsh, 'Schematism', p. 85.
26 Kant, *Critique of Pure Reason*, A51/B76.
27 Rae Langton, *Kantian Humility: Our Ignorance of Things in Themselves, Kantian Humility* (Oxford: Oxford University Press, 1998), p. 10.
28 Kant, *Prolegomena to Any Future Metaphysics*, 4: 292.
29 Kant, *Critique of Pure Reason*, A146/B186.
30 Thus, the schematism has an indirect relationship with the transcendental unity of apperception. Kant, *Critique of Pure Reason*, A145/B185. The transcendental unity of apperception is '[t]he I think which must be able to accompany all my representations'. Kant, *Critique of Pure Reason*, B132.
31 Sebastian Gardner, *Routledge Philosophy Guidebook to Kant and the Critique of Pure Reason* (London: Routledge, 1999), p. 171.
32 Whewell, *On the Philosophy of Discovery*, p. 312.
33 Kant, *Critique of Pure Reason*, A494/B523.
34 Kant, *Critique of Pure Reason*, Bxxii.
35 Whewell, *On the Philosophy of Discovery*, p. 313.
36 Immanuel Kant, 'Jäsche Logic', in J. Michael Young (ed.), *Lectures on Logic* (Cambridge: Cambridge University Press, 1992), 9: 131.
37 Kant, *Critique of Pure Reason*, A126.
38 Kant, *Critique of Pure Reason*, A80/B106.
39 Quentin Meillassoux, *After Finitude: An Essay on the Necessity of Contingency*, Ray Brassier (trans.) (London: Continuum, 2008), p. 118.
40 Whewell, *On the Philosophy of Discovery*, p. 314.
41 Whewell did not share Meillassoux's criticism of Kant as he commended Kant for revealing that knowledge necessarily required phenomena, whereas Meillassoux regards the limitation of knowledge to phenomena as the essential problem of transcendental idealism. Meillassoux is primarily concerned with demonstrating that Kant's philosophy requires absolute knowledge that goes beyond critical idealism. In short, he argues that the contingency inherent in all experience – whether entities remain the same or change at every moment – is the only absolute truth, which is necessary and eternal. In other words, '[t]he absolute is the absolute impossibility of a necessary being, or the necessity of contingency'. Meillassoux, *After Finitude*, p. 60.
42 Ducheyne argues that the late Kant became aware of the tension between metaphysics and physics within account and addressed it in the unpublished *Opus Postumum*. The *Metaphysical Foundations of Natural Science* isolated metaphysics from physics. Ducheyne describes the relation between metaphysics and physics as the 'bridging problem' and argues that this was source of Whewell's dissatisfaction with

Kant's philosophy. See Ducheyne, 'Kant and Whewell on Bridging Principles between Metaphysics and Science', 26.

43 Immanuel Kant, *Metaphysical Foundations of Natural Science*, Michael Friedman (ed. and trans.), Cambridge Texts in the History of Philosophy (Cambridge: Cambridge University Press, 2004), 4: 470.

44 James Kreines, 'Kant on the Laws of Nature: Laws, Necessitation, and the Limitation of Our Knowledge', *European Journal of Philosophy*, 17/4 (2009), 540.

45 Kreines, 'Kant on the Laws of Nature', 541. In this context, Kreines is opposing Freidman's suggestion that we have not yet discovered such particular laws to be indications that such laws could be discovered in principle.

46 Menachem Fisch, 'Necessary and Contingent Truth in William Whewell's Antithetical Theory of Knowledge', *Studies in History and Philosophy of Science Part A*, 16/4 (1985), 292.

47 Menachem Fisch, *William Whewell, Philosopher of Science* (Oxford: Clarendon Press, 1991), p. 105. For a similar account, Laura J. Snyder, *Reforming Philosophy: A Victorian Debate on Science and Society* (Chicago IL: University of Chicago Press, 2006), p. 46.

48 Steffen Ducheyne, 'Fundamental Questions and Some New Answers on Philosophical, Contextual and Scientific Whewell: Some Reflections on Recent Whewell Scholarship and the Progress Made Therein', *Perspectives on Science*, 18/2 (2010), 258.

49 Andrew Cooper, 'Reading Kant's *Kritik der Urteilskraft* in England, 1796–1840', *British Journal for the History of Philosophy* 29(3), 2021, 479.

50 Cf. Ducheyne, 'Fundamental Questions and Some New Answers on Philosophical, Contextual and Scientific Whewell', 255; Fisch, *William Whewell, Philosopher of Science*, p. 105.

51 Kant describes the thing in itself as an X that is 'nothing for us'. Kant, *Critique of Pure Reason*, A105. Whewell adopted this position in his unpublished notebooks as early as 1832. Cf. Ducheyne, 'Kant and Whewell on Bridging Principles between Metaphysics and Science', 22–45.

52 Whewell, *The Philosophy of the Inductive Sciences*, 1, p. 24.

53 Butts, *Historical Pragmatics*, p. 193.

54 Whewell, *The Philosophy of the Inductive Sciences*, 1, p. 42.

55 Whewell, *Novum Organon Renovatum*, p. 66.

56 Whewell, *The Philosophy of the Inductive Sciences*, 1, p. 65.

57 Whewell, *The Philosophy of the Inductive Sciences*, 1, p. 65.

58 Whewell, *On the Philosophy of Discovery*, p. 275.

59 Robert E. Butts (ed.), 'Induction as Unification: Kant, Whewell, and Recent Developments', in *Kant and Contemporary Epistemology* (Dordrecht: Springer Netherlands, 1994), p. 280.

60 Kant, *Critique of Pure Reason*, A663/B691.

61 Kant, *Prolegomena to Any Future Metaphysics*, 4: 332.

62 Kant, *Critique of Pure Reason*, A653/B681.

63 Allison, *Custom and Reason in Hume*, p. 143.

64 Rachel Zuckert, 'Empirical Scientific Investigation and the Ideas of Reason', in Angela Breitenbach and Michela Massimi (eds), *Kant and the Laws of Nature* (Cambridge: Cambridge University Press, 2017), p. 99.

65 Fisch, 'Necessary and Contingent Truth in William Whewell's Antithetical Theory of Knowledge', 309.

66 Whewell, *On the Philosophy of Discovery*, p. 344.

67 See Chapter 1 (section 1.1.2).

68 Kant, *Critique of Pure Reason*, A87/B119.

69 Whewell, *On the Philosophy of Discovery*, pp. 357–8.

70 Whewell, *On the Philosophy of Discovery*, pp. 358–9.

71 Whewell, *On the Philosophy of Discovery*, pp. 367–8.

72 Whewell, *On the Philosophy of Discovery*, p. 371.

73 Kant, *Critique of Pure Reason*, A494/B522.

74 Kant, *Critique of Pure Reason*, A598/B626.

75 Kant, *Critique of Pure Reason*, A600/B628.

76 Kant, *Critique of Pure Reason*, A601/B629.

77 Kant, *Critique of Pure Reason*, A763/B791.

78 Kant, *Critique of Pure Reason*, A679/B707.

79 Kant, *Metaphysical Foundations of Natural Science*, 4: 468.

80 Philip Kitcher, 'Projecting the Order of Nature', in Robert E. Butts (ed.), *Kant's Philosophy of Physical Science: Metaphysische Anfangsgründe Der Naturwissenschaft 1786–1986*, (Dordrecht: Springer Netherlands, 1986), p. 229.

81 Kitcher, 'Projecting the Order of Nature', p. 222.

82 Kant, *Critique of Pure Reason*, Axxi.

83 Kant, *Critique of Pure Reason*, A642/B670.

84 Henry E. Allison, *Kant's Transcendental Idealism* (London: Yale University Press, 2004), p. 430.

85 Kant, *Critique of Pure Reason*, A740-7/B768-75.

86 Kant, *Critique of Pure Reason*, A751/B779.

87 Angela Breitenbach and Yoon Choi, 'Pluralism and the Unity of Science', *The Monist*, 100/3 (2017), 396.

88 Breitenbach and Choi, 'Pluralism and the Unity of Science', 398–9.

89 Breitenbach and Choi, 'Pluralism and the Unity of Science', 401.

90 These included rewriting scientific textbooks after each 'paradigm shift' or directing future research towards auxiliary hypothesis rather the hardcore elements of scientific theory that would entail the refutation of the theory as a whole. See Chapter 1 (sections 1.2 and 1.3).

91 Kant, *Critique of Pure Reason*, A750/B778.
92 See Chapter 2 (section 2.3).
93 See Chapter 1 (section 1.4.3).
94 Angela Potochnik, *Idealization and the Aims of Science* (Chicago IL: University of Chicago Press, 2017), p. 213.
95 Potochnik, *Idealization and the Aims of Science*, p. 208.
96 John Mackie, *Ethics: Inventing Right and Wrong* (London: Penguin Books, 1990), p. 38.
97 John Mackie, 'A Refutation of Morals', *Australasian Journal of Psychology and Philosophy*, 24/1–2 (1946), 77.
98 Mackie, *Ethics*, p. 113.
99 Edward O. Wilson, *Sociobiology*, Abridged edition (Cambridge MA: Harvard University Press, 1980), p. 287.
100 Wilson, *Sociobiology*, p. 287.
101 This transcendental perspective does not seem to be limited to Kant's critical philosophy. Rather Wilson seems to regard transcendental to include any philosophical approach that does not recognise the importance of biological pressures.
102 Edward O. Wilson, *Consilience: The Unity of Knowledge* (New York: Random House, 1998), p. 263.
103 Wilson, *Consilience*, p. 282.
104 Whewell, *On the Philosophy of Discovery*, p. 164.
105 Larry Laudan, 'William Whewell on the Consilience of Inductions', *The Monist*, 55/3 (1971), 372.
106 Dupré, *Processes of Life*, p. 30.
107 John Dupré and Daniel J. Nicholson, 'A Manifesto for a Processual Philosophy of Biology', in *Everything Flows: Towards a Processual Philosophy of Biology* (Oxford: Oxford University Press, 2018), p. 4.

Chapter 4

1 William Whewell, *History of Scientific Ideas* (London: John W. Parker and Son, 1858), p. 197.
2 Whewell, *History of Scientific Ideas*, p. 240.
3 Kant, *Critique of the Power of Judgment*, 5: 410.
4 Kant, *Critique of the Power of Judgment*, 5: 387.
5 Kant, *Critique of the Power of Judgment*, 5: 387.
6 Philippe Huneman, 'Purposiveness, Necessity, and Contingency', in Ina Goy and Eric Watkins (eds), *Kant's Theory of Biology* (Berlin: De Gruyter, 2014), p. 192.
7 Whewell, *History of Scientific Ideas*, p. 251.
8 Lewontin, *The Triple Helix*, p. 3.

9 Whewell, *History of Scientific Ideas*, p. 252.
10 Whewell, *History of Scientific Ideas*, p. 254.
11 Whewell, *History of Scientific Ideas*, p. 253.
12 Kant, *Critique of the Power of Judgment*, 5: 368.
13 Kant, *Critique of the Power of Judgment*, 5: 368.
14 Richard Yeo, 'William Whewell, Natural Theology and the Philosophy of Science in Mid Nineteenth Century Britain', *Annals of Science*, 36/5 (1979), 498. For Whewell, the mind of humans can only comprehend the mind of God in a limited sense. Whewell compares the former to the shores of the sea and the latter to the ocean. See Chapter 3 (section 3.1.3).
15 This relates to the conception of influence developed in Chapter 1. I argued that influence should not be limited to the condition of similitude but should also allow for the possibility that influence can be based on misunderstanding or misprision. See Chapter 1 (section 1.4).
16 All references to Darwin's *Origin* correspond to the first edition (1859). Later editions contain less adequate expositions of Darwin's theory of natural selection. See editorial discussion on the text by Carroll in Charles Darwin, *On the Origin of Species*, Joseph Carroll (ed.) (Peterborough ON: Broadview Press, 2003), p. 76.
17 Darwin, *On the Origin of Species*, p. 379.
18 Ernst Mayr, *One Long Argument: Charles Darwin and the Genesis of Modern Evolutionary Thought* (Cambridge MA: Harvard University Press, 1991), pp. 36–7.
19 Elliott Sober, *Did Darwin Write the Origin Backwards? Philosophical Essays on Darwin's Theory* (Amherst NY: Prometheus Books, 2011), p. 28.
20 Darwin, *On the Origin of Species*, p. 373.
21 Darwin, *On the Origin of Species*, p. 360.
22 Sober, *Did Darwin Write the Origin Backwards?*, p. 44.
23 Snyder, *Reforming Philosophy*, p. 185.
24 Michael Ruse, 'Darwin's Debt to Philosophy: An Examination of the Influence of the Philosophical Ideas of John F.W. Herschel and William Whewell on the Development of Charles Darwin's Theory of Evolution', *Studies in History and Philosophy of Science Part A*, 6/2 (1975), 166.
25 John Herschel, *A Preliminary Discourse on the Study of Natural Philosophy* (London: Routledge/Thoemmes Press, 1996), p. 149.
26 Paul R. Thagard, 'Darwin and Whewell', *Studies in History and Philosophy of Science, Part A*, 8/4 (1977), 356.
27 Michael Ruse, 'Darwin and Herschel', *Studies in History and Philosophy of Science, Part A*, 9/4 (1978), 328.
28 Whewell denies that analogy could serve as the basis for *vera causa* principles because it would allow for entities such as eddying streams

to be evidence for the Cartesian hypothesis of vortices as a *vera causa*. Cf. Whewell, *The Philosophy of the Inductive Sciences*, 1, p. 283.

29 Charles H. Pence, 'Sir John F. W. Herschel and Charles Darwin: Nineteenth-Century Science and Its Methodology', *HOPOS: The Journal of the International Society for the History of Philosophy of Science*, 8/1 (2018), 122–3.

30 Darwin, *On the Origin of Species*, p. 378.

31 Kenneth Waters, 'The Arguments in the Origin of Species', in Jonathan Hodge and Gregory Radick (eds), *The Cambridge Companion to Darwin* (Cambridge: Cambridge University Press, 2003), p. 123.

32 Kant, *Critique of Pure Reason*, A653/B681.

33 Darwin, *On the Origin of Species*, p. 146.

34 Robert J. Richards, 'Darwin and the Inefficacy of Artificial Selection', *Studies in History and Philosophy of Science Part A*, 28/1 (1997), 77.

35 Richards, 'Darwin and the Inefficacy of Artificial Selection', 84.

36 Richards, 'Darwin and the Inefficacy of Artificial Selection', 95.

37 Ruse, 'Darwin's Debt to Philosophy', 343.

38 Darwin, *On the Origin of Species*, p. 134.

39 Darwin, *On the Origin of Species*, p. 157.

40 Darwin, *On the Origin of Species*, p. 162.

41 Ruse, 'Darwin's Debt to Philosophy', 350.

42 Michael Ruse, 'Charles Darwin and Artificial Selection', *Journal of the History of Ideas*, 36/2 (1975), 339–50.

43 Darwin, *On the Origin of Species*, p. 122.

44 The problem pertaining to offering a single definition of the organism is still an issue in contemporary philosophy of biology. This is discussed in Chapter 5 (section 5.1).

45 Darwin, *On the Origin of Species*, p. 162.

46 Kant, *Critique of the Power of Judgment*, 5: 360.

47 William Paley, *Natural Theology: Or, Evidences of the Existence and Attributes of the Deity, Collected from the Appearances of Nature*, 6th edition (Cambridge: Cambridge University Press, 2009), p. 2.

48 Paley, *Natural Theology*, p. 11.

49 Paley, *Natural Theology*, p. 12.

50 Francisco Ayala, 'From Paley to Darwin: Design to Natural Selection', in John B. Cobb Jr (ed.), *Back to Darwin: A Richer Account of Evolution* (Grand Rapids MI: William B Eerdmans Publishing Co., 2008), p. 70.

51 Kant, *Critique of the Power of Judgment*, 5: 370.

52 Kant, *Critique of the Power of Judgment*, 5: 374.

53 Clark Zumbach, *The Transcendent Science: Kant's Conception of Biological Methodology* (The Hague: Martinus Nijhoff Publishers, 1984), p. 4.

54 Hannah Ginsborg, *The Normativity of Nature: Essays on Kant's Critique of Judgement* (Oxford: Oxford University Press, 2015), p. 239.
55 Peter McLaughlin, *Kant's Critique of Teleology in Biological Explanation: Antinomy and Teleology* (Lewiston: Edwin Mellen Press Ltd, 1990), p. 178.
56 Ginsborg, *The Normativity of Nature*, p. 285.
57 McLaughlin, *Kant's Critique of Teleology in Biological Explanation*, p. 178.
58 McLaughlin, *Kant's Critique of Teleology in Biological Explanation*, p. 179.
59 Arthur Lovejoy, 'Kant and Evolution', in Bentley Glass, Owsei Temkin and William Straus Jr (eds), *Forerunners of Darwin, 1745–1859* (Baltimore MD: Johns Hopkins University Press, 1968), p. 175.
60 Michael Ruse, 'Reply to Richards', in *Debating Darwin* (Chicago IL: University of Chicago Press, 2016), p. 177.
61 Ruse, 'Reply to Richards', p. 199.
62 Michael Ruse, 'Charles Darwin: Great Briton', in *Debating Darwin* (Chicago IL: University of Chicago Press, 2016), p. 47.
63 Michael Ruse, *Darwin and Design: Does Evolution Have a Purpose?* (Cambridge MA: Harvard University Press, 2003), p. 284.
64 Robert J. Richards, 'Michael Ruse's Design for Living', *Journal of the History of Biology*, 37/1 (2004), 37.
65 This was also the basis for Huneman's claim that mechanical and teleological ways of conceiving nature were complementary to one another. See Chapter 4 (section 4.1.1).
66 Kant, *Critique of the Power of Judgment*, 5: 398.
67 Kant, *Critique of the Power of Judgment*, 5: 195
68 Paul Guyer, 'Organisms and the Unity of Science', in Eric Watkins (ed.), *Kant and the Sciences* (Oxford: Oxford University Press, 2001), p. 264.
69 Kant, *Critique of the Power of Judgment*, 5: 400.
70 In the next section, I will examine some arguments that have suggested that natural selection requires Kant's philosophy to explain the existence of organisms.
71 Richards, 'Michael Ruse's Design for Living', p. 33.
72 Kant, *Critique of the Power of Judgment*, 5: 445.
73 Kant, *Critique of the Power of Judgment*, 5: 450–1.
74 Kant, *Critique of the Power of Judgment*, 5: 456.
75 Robert J. Richards, 'Charles Darwin: Cosmopolitan Thinker', in *Debating Darwin* (Chicago IL: University of Chicago Press, 2016), p. 110.
76 Richards, 'Charles Darwin: Cosmopolitan Thinker', p. 127.
77 Darwin, *On the Origin of Species*, p. 146.

78 Richards, 'Charles Darwin: Cosmopolitan Thinker', p. 125.
79 Kant, *Critique of the Power of Judgment*, 5: 371.
80 Kant, *Critique of the Power of Judgment*, 5: 371.
81 This is a crucial aspect of Ginsborg's interpretation, which is further discussed in Chapter 5 (section 5.2.2). She describes the judgement of organisms as relating to a nought without a value to exemplify that it is not derived from rational principles and to show how defective organisms are not morally defective. Cf. Ginsborg, *The Normativity of Nature*, pp. 251–4.
82 Angela Breitenbach, 'Teleology in Biology: A Kantian Perspective', *Kant Yearbook*, 1/1 (2009), 45.
83 Ginsborg, *The Normativity of Nature*, p. 330.
84 Colin Allen and Marc Bekoff, 'Biological Function, Adaptation, and Natural Design', *Philosophy of Science*, 62/4 (1995), 612.
85 Ruse, *Darwin and Design*, p. 285.
86 Breitenbach, 'Teleology in Biology', 50.
87 Stephen. J. Gould and Richard C. Lewontin, 'The Spandrels of San Marco and the Panglossian Paradigm: A Critique of the Adaptationist Programme', *Proceedings of the Royal Society of London. Series B. Biological Sciences*, 205 (1979), 582.
88 Gould and Lewontin, 'The Spandrels of San Marco and the Panglossian Paradigm', 587.
89 Daniel C. Dennett, *Darwin's Dangerous Idea: Evolution and the Meanings of Life*, New Ed edition (London: Penguin, 1996), p. 248.
90 Stephen Jay Gould and Elisabeth S. Vrba, 'Exaptation – A Missing Term in the Science of Form', in David L. Hull and Michael Ruse (eds), *The Philosophy of Biology*, Oxford Readings in Philosophy (Oxford: Oxford University Press, 1998), p. 65.
91 Gould and Vrba, 'Exaptation – A Missing Term in the Science of Form', p. 67.
92 Gould and Vrba, 'Exaptation – A Missing Term in the Science of Form', p. 68.
93 Kant, *Critique of the Power of Judgment*, 5: 419.
94 Kant, *Critique of the Power of Judgment*, 5: 419–20.
95 This offers support to Mensch's account of how transcendental idealism appealed to the biological sciences, specifically the principles of epigenesis. However, these appeals to scientific principles to explain the emergence of reason ultimately entailed that these principles were not reconcilable with transcendental idealism in their biological context. According to Mensch, '[e]pigenesis thus served as a resource for a *metaphysical* portrait of reason, even as it was denied determinate efficacy in the world of organisms' Mensch, *Kant's Organicism*, p. 144. See Chapter 1 (section 1.1.2).

96 Zammito, 'Teleology Then and Now', 748.
97 Dennett, *Darwin's Dangerous Idea*, p. 50.
98 Matthew Ratcliffe, 'A Kantian Stance on the Intentional Stance', *Biology and Philosophy*, 16/1 (2001), 37–8.
99 Ratcliffe, 'A Kantian Stance on the Intentional Stance', 42.
100 Breitenbach, 'Teleology in Biology', 44.
101 Daniel J. Nicholson, 'The Concept of Mechanism in Biology', *Studies in History and Philosophy of Biological and Biomedical Sciences*, 43/1 (2012), 669.
102 Nicholson, 'The Concept of Mechanism in Biology', 671.
103 See Chapter 4 (section 4.2.2).
104 Ilya Prigogine and Isabelle Stengers, *Order Out of Chaos: Man's New Dialogue with Nature* (New York: Bantam Books, 1984), p. 115.
105 Erwin Schrödinger, *What is Life?: With Mind and Matter and Autobiographical Sketches* (Cambridge: Cambridge University Press, 2012), p. 77.
106 Daniel J. Nicholson, 'Reconceptualizing the Organism: From Complex Machine to Flowing Stream', in *Everything Flows: Towards a Processual Philosophy of Biology* (Oxford: Oxford University Press, 2018), p. 146.
107 John Dupré and Stephan Guttinger, 'Viruses as Living Processes', *Studies in History and Philosophy of Biological and Biomedical Sciences*, 59 (2016), 110.
108 Nicholson, 'Reconceptualizing the Organism', 159.
109 Stuart A. Kauffman, *At Home in the Universe: The Search for Laws of Self-Organization and Complexity* (London: Penguin, 1996), pp. 20–1.
110 Ginsborg, *The Normativity of Nature*, p. 344.
111 Ginsborg, *The Normativity of Nature*, p. 320.
112 Alicia Juarrero Roqué, 'Self-Organization: Kant's Concept of Teleology and Modern Chemistry', *The Review of Metaphysics*, 39/1 (1985), 108–9.
113 Andreas Weber and Francisco J. Varela, 'Life after Kant: Natural Purposes and the Autopoietic Foundations of Biological Individuality', *Phenomenology and the Cognitive Sciences*, 1/2 (2002), 121.

Chapter 5

1 Samir Okasha and Ellen Clarke, 'Species and Organisms: What Are the Problems?', in Frédéric Bouchard and Philippe Huneman (eds), *From Groups to Individuals: Evolution and Emerging Individuality* (Cambridge MA: MIT Press, 2013), p. 55.
2 Darwin, *On the Origin of Species*, p. 122.
3 Frédéric Bouchard, 'Causal Processes, Fitness, and the Differential Persistence of Lineages', *Philosophy of Science*, 75/5 (2008), 633.

4 Ellen Clarke, 'The Problem of Biological Individuality', *Biological Theory*, 5/4 (2010), 315–16.

5 John W. Pepper and Matthew D. Herron, 'Does Biology Need an Organism Concept?', *Biological Reviews*, 83/4 (2008), 624.

6 Michael Grant and Jeffry Mitton, 'Case Study: The Glorious, Golden, and Gigantic Quaking Aspen', *Nature Education Knowledge*, 3/10 (2010), 40.

7 Frédéric Bouchard, 'Symbiosis, Lateral Function Transfer and the (Many) Saplings of Life', *Biology & Philosophy*, 25/4 (2010), 632.

8 Lora V. Hooper, 'Bacterial Contributions to Mammalian Gut Development', *Trends in Microbiology*, 12/3 (2004), 130.

9 Peter J. Turnbaugh and others, 'The Human Microbiome Project: Exploring the Microbial Part of Ourselves in a Changing World', *Nature*, 449/7164 (2007), 804–10.

10 The recognition of the need for a greater understanding between biological entities and their environments has been the focus of a movement called 'niche construction'. For an account of the significance of niche construction for philosophy of biology, see F. John Odling-Smee, Kevin Laland and Marcus Feldman, *Niche Construction: The Neglected Process in Evolution* (Princeton NJ: Princeton University Press, 2003).

11 Scott Turner, 'Superorganisms and Superindividuality: The Emergence of Individuality in a Social Insect Assemblage', in Frédéric Bouchard and Philippe Huneman (eds), *From Groups to Individuals: Evolution and Emerging Individuality* (Cambridge MA: The MIT Press, 2013), p. 224.

12 Richard Dawkins, *The Extended Phenotype: The Long Reach of the Gene*, Revised edition (Oxford: Oxford University Press, 1999).

13 Turner, 'Superorganisms and Superindividuality', p. 236.

14 J. Scott Turner, 'Extended Phenotypes and Extended Organisms', *Biology and Philosophy*, 19/3 (2004), 347.

15 Bouchard, 'Causal Processes, Fitness, and the Differential Persistence of Lineages', 567.

16 Turner, 'Extended Phenotypes and Extended Organisms', 345.

17 See Chapter 4 (section 4.4.2).

18 John Symons, 'The Individuality of Artifacts and Organisms', *History and Philosophy of the Life Sciences*, 32/2/3 (2010), 245.

19 See Chapter 4 (section 4.1.1).

20 Symons, 'The Individuality of Artifacts and Organisms', 244.

21 Bouchard, 'Symbiosis, Lateral Function Transfer and the (Many) Saplings of Life', 631.

22 Frédéric Bouchard, 'What Is a Symbiotic Superindividual and How Do You Measure Its Fitness?', in Frédéric Bouchard and Philippe

Huneman (eds), *From Groups to Individuals: Evolution and Emerging Individuality* (Cambridge MA: MIT Press, 2013), p. 244.

23 Stuart A. Kauffman, 'Evolution Beyond Newton, Darwin, and Entailing Law', in *Beyond Mechanism: Putting Life Back into Biology* (Plymouth: Lexington Books, 2013), p. 5.

24 Kauffman, p. 68.

25 Alvaro Moreno and Matteo Mossio, *Biological Autonomy: A Philosophical and Theoretical Enquiry*, History, Philosophy and Theory of the Life Sciences (Dordrecht: Springer Netherlands, 2015), p. 149.

26 Moreno and Mossio, *Biological Autonomy*, p. 149. The role of the analogy between artefacts and organisms for Darwin's theory was discussed in Chapter 4 (section 4.2).

27 Stuart Kauffman and Philip Clayton, 'On Emergence, Agency, and Organization', *Biology & Philosophy*, 21/4 (2006), 510.

28 Matteo Mossio and Alvaro Moreno, 'Organisational Closure in Biological Organisms', *History and Philosophy of the Life Sciences*, 32/2–3 (2010), 278.

29 Moreno and Mossio, *Biological Autonomy*, p. 17.

30 Moreno and Mossio, *Biological Autonomy*, p. 18.

31 Nicholson, 'Reconceptualizing the Organism', p. 152.

32 Moreno and Mossio, *Biological Autonomy*, p. 61.

33 Moreno and Mossio, *Biological Autonomy*, p. 58.

34 Kant, *Critique of the Power of Judgment*, 5: 444.

35 Kant, *Critique of the Power of Judgment*, 5: 445.

36 Ginsborg, *The Normativity of Nature*, p. 252.

37 Ginsborg, *The Normativity of Nature*, p. 244.

38 Kant, *Critique of the Power of Judgment*, 5: 479.

39 Angela Breitenbach, 'Biological Purposiveness and Analogical Reflection', in Eric Watkins and Ina Goy (eds), *Kant's Theory of Biology* (Berlin: De Gruyter, 2014), p. 136fn.7.

40 Kant, *Critique of the Power of Judgment*, 5: 480.

41 Kant, *Critique of the Power of Judgment*, 5: 481.

42 Kant, *Critique of the Power of Judgment*, 5: 478.

43 Kant, *Critique of the Power of Judgment*, 5: 479.

44 Clarke, 'The Problem of Biological Individuality', 313.

45 Peter Godfrey-Smith, *Philosophy of Biology* (Princeton NJ: Princeton University Press, 2014), p. 68.

46 Dupré, *Human Nature and the Limits of Science*, p. 179.

47 Dupré, *Processes of Life*, p. 282. In relation to the various conceptions of biological individuality discussed in the first section of this chapter, Dupré's account is similar to Turner's account of the extended organism insofar as both suggest that the structures that biological entities construct from their environment should be regarded as relevant for

our understanding of biological entities. There are two important differences between their accounts. First, for Turner, the structures that organisms construct, such as termite mounds, are extensions of the physiology of the organism. Second, Turner emphasises that his account is still compatible with a genetic account of the organism, albeit as inclusive of the extended physiologies that are part of that organism.

48 Dupré, *Processes of Life*, p. 291.
49 Dupré, *Processes of Life*, p. 291
50 Dupré, *Processes of Life*, p. 291.
51 Kant, *Critique of Pure Reason*, A578-9/B606-7.
52 Kant, *Critique of Pure Reason*, A580/B608. This relates to Chapter 2 where I argued that Kant's conception of natural laws, as regulative principles that are transcendental conditions for possibility of science, provides a middle ground between the accounts of laws developed by Bhaskar and Cartwright. Kant is in agreement with Cartwright and Dupré insofar as he denies the ontological necessity of natural laws independent of experience for broadly empiricist reasons, yet he argues that this does not refute the need for a regulative conception of laws for the possibility of scientific enquiry. In alignment with Bhaskar's account, these laws are necessary for the possibility of science; however, the point of divergence between Bhaskar and Kant is marked by Bhaskar's argument that this justifies the ontological status of such laws. See Chapter 2 (section 2.3).
53 Kant, *Critique of Pure Reason*, A192/B237.
54 Kant, *Critique of Pure Reason*, A203/B248.
55 John Dupré, *The Disorder of Things: Metaphysical Foundations of the Disunity of Science* (Cambridge MA: Harvard University Press, 1993), pp. 176–7.
56 Kant, *Critique of Pure Reason*, A127.
57 Allison, *Kant's Transcendental Idealism*, pp. 425–30.
58 See Chapter 3 (section 3.1.1).
59 Kant, *Critique of Pure Reason*, A644/B672.
60 Kant, *Critique of Pure Reason*, A336/B564.
61 Kant, 'Critique of Practical Reason', 5: 97.
62 See Chapter 2 (section 2.2.2).
63 Kant, 'Critique of Practical Reason', 5: 94.
64 Wood, 'Kant's Compatibilism', p. 74.
65 Kant, 'Critique of Practical Reason', 5: 134.
66 Kant, *Groundwork of the Metaphysics of Morals*, 4: 414.
67 David L. Hull, *The Metaphysics of Evolution* (New York: State University of New York Press, 1989), p. 2.
68 Hull, *The Metaphysics of Evolution*, p. 12.

69 Kant, *Groundwork of the Metaphysics of Morals*, 4: 438.
70 Kant, *Groundwork of the Metaphysics of Morals*, 4: 433.
71 Kant, *Groundwork of the Metaphysics of Morals*, 4: 411.
72 Kant, *Groundwork of the Metaphysics of Morals*, 4: 414.
73 Kant, 'Critique of Practical Reason', 5: 65.
74 Hull, *The Metaphysics of Evolution*, pp. 23–4.
75 See Chapter 3 (section 3.1).
76 This is broadly compatible with the commitments to consilience previously discussed in relation to Mackie and Wilson. They also suggested any account of morality must be compatible with biology. See Chapter 3 (section 3.2.2).
77 Kant, *Critique of Pure Reason*, A645/B673.
78 Kant, *Critique of Pure Reason*, A651/B679.
79 See Chapter 3 (section 3.2.1).
80 Kant, *Groundwork of the Metaphysics of Morals*, 4: 421.
81 See Chapter 5 (section 5.2.2).
82 Kant, *Groundwork of the Metaphysics of Morals*, 4: 397.
83 Wood, 'Kant's Compatibilism', p. 99.
84 Dupré, *Processes of Life*, p. 284.
85 Kant, 'The Metaphysics of Morals', 6: 326.
86 Kant, 'The Metaphysics of Morals', 6: 326.
87 A related aspect of Kant's political philosophy was discussed in Chapter 1 (section 1.2.2). I compared Kant's opposition to political revolutions with Kuhn's conception of scientific revolutions and concluded that Kant would have opposed Kuhn's appeal to political revolutions as support for scientific revolutions. For Kant, political revolutions are not compatible with the development of society in accordance with the demands of reason.
88 Kant, 'The Metaphysics of Morals', 6: 318.
89 Kant, 'The Metaphysics of Morals', 6: 217.
90 Kant, 'The Metaphysics of Morals', 6: 216.
91 See Chapter 3 (section 3.2.2).
92 Christine M. Korsgaard, *The Sources of Normativity* (Cambridge: Cambridge University Press, 1996), p. 166.
93 Dupré, *Processes of Life*, p. 58.
94 Dupré, *Processes of Life*, p. 59.
95 Dupré, *Human Nature and the Limits of Science*, p. 63.
96 Dupré, *Processes of Life*, p. 60.

Bibliography

Allen, Colin, and Marc Bekoff, 'Biological Function, Adaptation, and Natural Design', *Philosophy of Science*, 62/4 (1995), 609–22.

Allison, Henry E., *Custom and Reason in Hume: A Kantian Reading of the First Book of the Treatise, Custom and Reason in Hume* (Oxford: Oxford University Press, 2008).

— *Idealism and Freedom: Essays on Kant's Theoretical and Practical Philosophy* (Cambridge: Cambridge University Press, 1996).

— *Kant's Theory of Freedom* (Cambridge: Cambridge University Press, 1990).

— *Kant's Transcendental Idealism* (London: Yale University Press, 2004).

Ayala, Francisco, 'From Paley to Darwin: Design to Natural Selection', in John B. Cobb Jr (ed.), *Back to Darwin: A Richer Account of Evolution* (Grand Rapids MI: William B. Eerdmans Publishing Co., 2008), pp. 50–75.

Beck, Lewis White, 'A Prussian Hume and a Scottish Kant', in *Essays on Kant and Hume* (London: Yale University Press, 1978), pp. 111–29.

Bhaskar, Roy, *A Realist Theory of Science* (London: Verso, 2008).

— *Reclaiming Reality: A Critical Introduction to Contemporary Philosophy* (London: Verso, 1989).

Bloom, Harold, *The Anxiety of Influence: A Theory of Poetry*, 2nd edition (New York: Oxford University Press, 1997).

Boer, Karin de, 'Kant's Response to Hume's Critique of Pure Reason', *Archiv für Geschichte der Philosophie*, 101/3 (2019), 376–406.

Bouchard, Frédéric, 'Causal Processes, Fitness, and the Differential Persistence of Lineages', *Philosophy of Science*, 75/5 (2008), 560–70.

— 'Symbiosis, Lateral Function Transfer and the (Many) Saplings of Life', *Biology & Philosophy*, 25/4 (2010), 623–41.

— 'What Is a Symbiotic Superindividual and How Do You Measure Its Fitness?', in Frédéric Bouchard and Philippe Huneman (eds), *From Groups to Individuals: Evolution and Emerging Individuality* (Cambridge MA: MIT Press, 2013), pp. 243–64.

Breitenbach, Angela, 'Biological Purposiveness and Analogical Reflection', in Eric Watkins and Ina Goy (eds), *Kant's Theory of Biology* (Berlin: De Gruyter, 2014), pp. 131–48.

— 'Teleology in Biology: A Kantian Perspective', *Kant Yearbook*, 1/1 (2009), 31–56.

Breitenbach, Angela, and Yoon Choi, 'Pluralism and the Unity of Science', *The Monist*, 100/3 (2017), 391–405.

Butts, Robert E., *Historical Pragmatics: Philosophical Essays* (Dordrecht: Kluwer Academic Publishers, 1993).

— 'Induction as Unification: Kant, Whewell, and Recent Developments', in Paolo Parrini (ed.), *Kant and Contemporary Epistemology* (Dordrecht: Springer Netherlands, 1994), pp. 273–89.

Cartwright, Nancy, *The Dappled World: A Study of the Boundaries of Science* (Cambridge: Cambridge University Press, 1999).

Clarke, Ellen, 'The Problem of Biological Individuality', *Biological Theory*, 5/4 (2010), 312–25.

Clarke, Stephen, 'Transcendental Realisms in the Philosophy of Science: On Bhaskar and Cartwright', *Synthese*, 173/3 (2010), 299–315.

Coleman, William, *Biology in the Nineteenth Century: Problems of Form, Function and Transformation* (Cambridge: Cambridge University Press, 1977).

Collingwood, Robin G., *An Essay on Metaphysics* (Oxford: Clarendon Press, 1940).

Cooper, Andrew, 'Reading Kant's *Kritik der Urteilskraft* in England, 1796–1840', *British Journal for the History of Philosophy*, 29/3 (2021), 472–93.

Darwin, Charles, *On the Origin of Species*, Joseph Carroll (ed.), 1st edition (Peterborough ON: Broadview Press, 2003).

Dawkins, Richard, *The Extended Phenotype: The Long Reach of the Gene*, Revised edition (Oxford: Oxford University Press, 1999).

Dennett, Daniel C., *Darwin's Dangerous Idea: Evolution and the Meanings of Life* (London: Penguin, 1996).

Ducheyne, Steffen, 'Fundamental Questions and Some New Answers on Philosophical, Contextual and Scientific Whewell: Some Reflections on Recent Whewell Scholarship and the Progress Made Therein', *Perspectives on Science*, 18/2 (2010), 242–72.

— 'Kant and Whewell on Bridging Principles between Metaphysics and Science', *Kant-Studien*, 102/1 (2011), 22–45.

Dupré, John, *Human Nature and the Limits of Science* (Oxford: Oxford University Press, 2001).

— *Processes of Life: Essays in the Philosophy of Biology* (Oxford: Oxford University Press, 2012).

— *The Disorder of Things: Metaphysical Foundations of the Disunity of Science* (Cambridge MA: Harvard University Press, 1993).

Dupré, John, and Stephan Guttinger, 'Viruses as Living Processes', *Studies in History and Philosophy of Biological and Biomedical Sciences*, 59 (2016), 109–16.

Dupré, John, and Daniel J. Nicholson, 'A Manifesto for a Processual Philosophy of Biology', in Daniel J. Nicholson and John Dupré (eds), *Everything Flows: Towards a Processual Philosophy of Biology* (Oxford: Oxford University Press, 2018), pp. 3–45.

Engstrom, Stephen, 'Knowledge and Its Object', in James R. O'Shea (ed.), *Kant's Critique of Pure Reason: A Critical Guide* (Cambridge: Cambridge University Press, 2017), pp. 28–45.

Fisch, Menachem, 'Necessary and Contingent Truth in William Whewell's Antithetical Theory of Knowledge', *Studies in History and Philosophy of Science Part A*, 16/4 (1985), 275–314.

— *William Whewell, Philosopher of Science* (Oxford: Clarendon Press, 1991).

Feyerabend, Paul. K., *Against Method* (London: Verso, 1978).

Floyd, Juliet, 'The Fact of Judgment: The Kantian Response to the Humean Condition', in Jeff Malpas (ed.), *From Kant to Davidson: Philosophy and the Idea of the Transcendental* (London: Routledge, 2003), pp. 22–47.

Gardner, Sebastian, *Routledge Philosophy Guidebook to Kant and the Critique of Pure Reason* (London: Routledge, 1999).

Ginsborg, Hannah, *The Normativity of Nature: Essays on Kant's Critique of Judgement* (Oxford: Oxford University Press, 2015).

Godfrey-Smith, Peter, *Philosophy of Biology* (Princeton NJ: Princeton University Press, 2014).

Gotterbarn, Donald, 'Kant, Hume and Analyticity', *Kant-Studien*, 65/1–4 (1974), 274–83.

Gould, Stephen. J., and Richard. C. Lewontin, 'The Spandrels of San Marco and the Panglossian Paradigm: A Critique of the Adaptationist Programme', *Proceedings of the Royal Society of London. Series B. Biological Sciences*, 205 (1979), 581–98.

Gould, Stephen Jay, and Elisabeth S. Vrba, 'Exaptation – A Missing Term in the Science of Form', in David L. Hull and Michael Ruse (eds), *The Philosophy of Biology* (Oxford: Oxford University Press, 1998), pp. 52–71.

Grant, Michael, and Jeffry Mitton, 'Case Study: The Glorious, Golden, and Gigantic Quaking Aspen', *Nature Education Knowledge*, 3/10 (2010), 40.

Grene, Marjorie, and David Depew, *The Philosophy of Biology: An Episodic History* (Cambridge: Cambridge University Press, 2004).

Guyer, Paul, 'Imperfect Knowledge of Nature', in Angela Breitenbach and Michela Massimi (eds), *Kant and the Laws of Nature* (Cambridge: Cambridge University Press, 2017), 49–68.

— *Kant* (New York: Routledge, 2006).

— *Knowledge, Reason, and Taste: Kant's Response to Hume* (Princeton NJ: Princeton University Press, 2008).

— 'Organisms and the Unity of Science', in Eric Watkins (eds), *Kant and the Sciences* (Oxford: Oxford University Press, 2001), pp. 259–82.

Herschel, John, *A Preliminary Discourse on the Study of Natural Philosophy* (London: Routledge/Thoemmes Press, 1996).

Hooper, Lora V., 'Bacterial Contributions to Mammalian Gut Development', *Trends in Microbiology*, 12/3 (2004), 129–34.

Hull, David L., *The Metaphysics of Evolution* (New York: State University of New York Press, 1989).

Hume, David, *A Treatise of Human Nature*, L. A. Selby-Bigge and P. H. Nidditch (eds), 2nd edition (Oxford: Oxford University Press, 1978).

—, in L. A. Selby-Bigge and P. H. Nidditch (eds), *Enquiries Concerning Human Understanding and Concerning the Principles of Morals*, 3rd edition (Oxford: Oxford University Press, 1975).

Huneman, Philippe, 'Purposiveness, Necessity, and Contingency', in Ina Goy and Eric Watkins (eds), *Kant's Theory of Biology* (Berlin: De Gruyter, 2014), pp. 185–202.

Kant, Immanuel, *Correspondence*, Arnulf Zweig (ed.) (Cambridge: Cambridge University Press, 1999).

— 'Critique of Practical Reason', in Mary J. Gregor (eds), *Practical Philosophy* (Cambridge: Cambridge University Press, 1996), pp. 133–272.

— *Critique of Pure Reason*, Paul Guyer and Allen W. Wood (trans.) (Cambridge: Cambridge University Press, 1998).

— *Critique of the Power of Judgment*, Paul Guyer (ed.), Paul Guyer and Eric Matthews (trans.) (Cambridge: Cambridge University Press, 2000).

— *Groundwork of the Metaphysics of Morals*, Mary Gregor (ed.), Christine M. Korsgaard (trans.) (Cambridge: Cambridge University Press, 1998).

— 'Jäsche Logic', in J. Michael Young (ed.), *Lectures on Logic* (Cambridge: Cambridge University Press, 1992), pp. 517–640.

— *Metaphysical Foundations of Natural Science*, Michael Friedman (ed. and trans.) (Cambridge: Cambridge University Press, 2004).

— *Prolegomena to Any Future Metaphysics: With Selections from the Critique of Pure Reason*, Gary Hatfield (ed.) (Cambridge: Cambridge University Press, 1997).

— 'The Metaphysics of Morals', in Mary J. Gregor (ed.), *Practical Philosophy* (Cambridge: Cambridge University Press, 1996), pp. 353–604.

— 'What Real Progress Has Metaphysics Made in Germany since the Time of Leibniz and Wolff?', in Gary Hatfield and Henry Allison (eds), *Theoretical Philosophy after 1781*, Michael Friedman and Peter Heath (trans.) (Cambridge: Cambridge University Press, 2002), pp. 337–424.

Kauffman, Stuart A., *At Home in the Universe: The Search for Laws of Self-Organization and Complexity* (London: Penguin, 1996).

— 'Evolution Beyond Newton, Darwin, and Entailing Law', in *Beyond Mechanism: Putting Life Back into Biology* (Plymouth: Lexington Books, 2013), pp. 1–24.

— *Investigations* (Oxford: Oxford University Press, 2000).

Kauffman, Stuart, and Philip Clayton, 'On Emergence, Agency, and Organization', *Biology & Philosophy*, 21/4 (2006), 501–21.

Kuhn, Thomas, *The Structure of Scientific Revolutions*, Enlarged edition (Chicago IL: Chicago University Press, 1970).

Kitcher, Philip, 'Projecting the Order of Nature', in Robert E. Butts (ed.), *Kant's Philosophy of Physical Science: Metaphysische Anfangsgründe Der Naturwissenschaft 1786–1986*, (Dordrecht: Springer Netherlands, 1986), pp. 201–35.

Korsgaard, Christine M., *The Sources of Normativity* (Cambridge: Cambridge University Press, 1996).

Kreines, James, 'Kant on the Laws of Nature: Laws, Necessitation, and the Limitation of Our Knowledge', *European Journal of Philosophy*, 17/4 (2009), 527–58.

Kuehn, Manfred, *Kant: A Biography* (Cambridge: Cambridge University Press, 2001).

— 'Kant's Conception of "Hume's Problem"', *Journal of the History of Philosophy*, 21/2 (1983), 175–93.

Langton, Rae, *Kantian Humility: Our Ignorance of Things in Themselves*, *Kantian Humility* (Oxford: Oxford University Press, 1998).

Laudan, Larry, 'William Whewell on the Consilience of Inductions', *The Monist*, 55/3 (1971), 368–91.

Leibniz, Gottfried Wilhelm, *Philosophical Essays*, Roger Ariew and Daniel Garber (eds) (Indianapolis: Hackett, 1989).

Lewontin, Richard, *The Triple Helix: Gene, Organism and Environment* (Cambridge MA: Harvard University Press, 2000).

Lidgard, Scott, and Lynn K. Nyhart, 'The Work of Biological Individuality: Concepts and Contexts', in Scott Lidgard and Lynn K. Nyhart (eds), *Biological Individuality: Integrating Scientific, Philosophical, and Historical Perspectives* (Chicago IL: University of Chicago Press, 2017), pp. 17–62.

Lovejoy, Arthur, 'Kant and Evolution', in Bentley Glass, Owsei Temkin and William Straus Jr (eds), *Forerunners of Darwin, 1745–1859* (Baltimore MD: Johns Hopkins University Press, 1968), pp. 173–206.

Mackie, John, 'A Refutation of Morals', *Australasian Journal of Psychology and Philosophy*, 24/1–2 (1946), 77–90.

— *Ethics: Inventing Right and Wrong* (London: Penguin Books, 1990).

Maliks, Reidar, *Kant's Politics in Context*, Kant's Politics in Context (Oxford: Oxford University Press, 2014).

Mayr, Ernst, *One Long Argument: Charles Darwin and the Genesis of Modern Evolutionary Thought* (Cambridge MA: Harvard University Press, 1991).

McLaughlin, Peter, *Kant's Critique of Teleology in Biological Explanation: Antinomy and Teleology* (Lewiston NY: Edwin Mellen Press Ltd, 1990).

Meillassoux, Quentin, *After Finitude: An Essay on the Necessity of Contingency*, Ray Brassier (trans.) (London: Continuum, 2008).

Mensch, Jennifer, *Kant's Organicism: Epigenesis and the Development of Critical Philosophy* (Chicago IL: University of Chicago Press, 2013).

Moreno, Alvaro, and Matteo Mossio, *Biological Autonomy: A Philosophical and Theoretical Enquiry*, History, Philosophy and Theory of the Life Sciences (Dordrecht: Springer Netherlands, 2015).

Mossio, Matteo, and Alvaro Moreno, 'Organisational Closure in Biological Organisms', *History and Philosophy of the Life Sciences*, 32/2–3 (2010), 269–88.

Nicholson, Daniel J., 'Reconceptualizing the Organism: From Complex Machine to Flowing Stream', in Daniel J. Nicholson and John Dupré (eds), *Everything Flows: Towards a Processual Philosophy of Biology* (Oxford: Oxford University Press, 2018), pp. 139–66.

— 'The Concept of Mechanism in Biology', *Studies in History and Philosophy of Biological and Biomedical Sciences*, 43/1 (2012), 152–63.

Odling-Smee, F. John, Kevin Laland and Marcus Feldman, *Niche Construction: The Neglected Process in Evolution* (Princeton NJ: Princeton University Press, 2003).

Okasha, Samir, and Ellen Clarke, 'Species and Organisms: What Are the Problems?', in Frédéric Bouchard and Philippe Huneman (eds), *From Groups to Individuals: Evolution and Emerging Individuality* (Cambridge MA: MIT Press, 2013), pp. 55–76.

O'Shea, James R., *Kant's Critique of Pure Reason: An Introduction and Interpretation* (Durham: Acumen Publishing, 2012).

Paley, William, *Natural Theology: Or, Evidences of the Existence and Attributes of the Deity, Collected from the Appearances of Nature*, 6th edition (Cambridge: Cambridge University Press, 2009).

Pence, Charles H., 'Sir John F. W. Herschel and Charles Darwin: Nineteenth-Century Science and Its Methodology', *HOPOS: The Journal of the International Society for the History of Philosophy of Science*, 8/1 (2018), 108–40.

Pepper, John W., and Matthew D. Herron, 'Does Biology Need an Organism Concept?', *Biological Reviews*, 83/4 (2008), 621–7.

Popper, Karl R, *The Logic of Scientific Discovery* (London: Routledge, 2002).

Potochnik, Angela, *Idealization and the Aims of Science* (Chicago IL: University of Chicago Press, 2017).

Prigogine, Ilya, and Isabelle Stengers, *Order Out of Chaos: Man's New Dialogue with Nature* (New York: Bantam Books, 1984).

Ratcliffe, Matthew, 'A Kantian Stance on the Intentional Stance', *Biology and Philosophy*, 16/1 (2001), 29–52.

Reichenbach, Hans, *Experience and Prediction: An Analysis of the Foundations and the Structure of Knowledge* (Chicago IL: University of Chicago Press, 1938).

Richards, Robert J., 'Charles Darwin: Cosmopolitan Thinker', in *Debating Darwin* (Chicago IL: University of Chicago Press, 2016), pp. 83–150.

— 'Darwin and the Inefficacy of Artificial Selection', *Studies in History and Philosophy of Science Part A*, 28/1 (1997), 75–97.

— 'Michael Ruse's Design for Living', *Journal of the History of Biology*, 37/1 (2004), 25–38.

Richman, Kenneth, 'Introduction', in Rupert Read and Kenneth Richman (eds), *The New Hume Debate*, Revised edition (London: Routledge, 2007), 1–15.

Roqué, Alicia Juarrero, 'Self-Organization: Kant's Concept of Teleology and Modern Chemistry', *The Review of Metaphysics*, 39/1 (1985), 107–35.

Ruse, Michael, 'Charles Darwin and Artificial Selection', *Journal of the History of Ideas*, 36/2 (1975), 339–50.

— 'Charles Darwin: Great Briton', in *Debating Darwin* (Chicago IL: University of Chicago Press, 2016), pp. 1–81.

— *Darwin and Design: Does Evolution Have a Purpose?* (Cambridge MA: Harvard University Press, 2003).

— 'Darwin and Herschel', *Studies in History and Philosophy of Science Part A*, 9/4 (1978), 323–31.

— 'Darwin's Debt to Philosophy: An Examination of the Influence of the Philosophical Ideas of John F. W. Herschel and William Whewell on the Development of Charles Darwin's Theory of Evolution', *Studies in History and Philosophy of Science Part A*, 6/2 (1975), 159–81.

— 'Reply to Richards', in *Debating Darwin* (Chicago IL: University of Chicago Press, 2016), pp. 177–202.

Schrödinger, Erwin, *What Is Life?: With Mind and Matter and Autobiographical Sketches*, Canto Classics (Cambridge: Cambridge University Press, 2012).

Snyder, Laura J., *Reforming Philosophy: A Victorian Debate on Science and Society* (Chicago IL: University of Chicago Press, 2006).

Sober, Elliott, *Did Darwin Write the Origin Backwards? Philosophical Essays on Darwin's Theory* (Amherst NY: Prometheus Books, 2011).

Stern, Robert, 'Transcendental Arguments: A Plea for Modesty', *Grazer Philosophische Studien*, 74/1 (2007), 143–61.

Strawson, Galen, 'David Hume: Objects and Powers', in *Real Materialism: And Other Essays* (Oxford: Oxford University Press, 2008), pp. 415–38.

— *The Secret Connexion: Causation, Realism, and David Hume*, Revised edition (Oxford: Oxford University Press, 2014).

Strawson, Peter, *The Bounds of Sense: An Essay on Kant's Critique of Pure Reason* (London: Methuen and Co., 1966).

Stroud, Barry, 'The Charm of Naturalism', *Proceedings and Addresses of the American Philosophical Association*, 70/2 (1996), 43–55.

— 'Transcendental Arguments', *The Journal of Philosophy*, 65/9 (1968), 241–56.

Symons, John, 'The Individuality of Artifacts and Organisms', *History and Philosophy of the Life Sciences*, 32/2/3 (2010), 233–46.

Thagard, Paul R., 'Darwin and Whewell', *Studies in History and Philosophy of Science Part A*, 8/4 (1977), 353–6.

Turnbaugh, Peter J., Ruth E. Ley, Micah Hamady, Claire Fraser-Liggett, Rob Knight and Jeffrey I. Gordon, 'The Human Microbiome Project: Exploring the Microbial Part of Ourselves in a Changing World', *Nature*, 449/7164 (2007), 804–10.

Turner, J. Scott, 'Extended Phenotypes and Extended Organisms', *Biology and Philosophy*, 19/3 (2004), 327–52.

Turner, Scott, 'Superorganisms and Superindividuality: The Emergence of Individuality in a Social Insect Assemblage', in Frédéric Bouchard and Philippe Huneman (eds), *From Groups to Individuals: Evolution and Emerging Individuality* (Cambridge MA: The MIT Press, 2013), pp. 219–42.

Walsh, W. H., 'Schematism', in *Kant: A Collection of Critical Essays*, Robert Paul Wolff (ed.) (New York: Anchor Books, 1967), pp. 71–87.

Waters, Kenneth, 'The Arguments in the Origin of Species', in Jonathan Hodge and Gregory Radick (eds), *The Cambridge Companion to Darwin* (Cambridge: Cambridge University Press, 2003), pp. 116–40.

Watkins, Eric, *Kant and the Metaphysics of Causality* (Cambridge: Cambridge University Press, 2005).

— 'Kant on the Distinction between Sensibility and Understanding', in James R. O'Shea (ed.), *Kant's Critique of Pure Reason: A Critical Guide* (Cambridge: Cambridge University Press, 2017), pp. 9–27.

Weber, Andreas, and Francisco J. Varela, 'Life after Kant: Natural Purposes and the Autopoietic Foundations of Biological Individuality', *Phenomenology and the Cognitive Sciences*, 1/2 (2002), 97–125.

'What Does Anschauung Mean?', *The Monist*, 2/4 (1892), 527–32.

Whewell, William, *History of Scientific Ideas* (London: John W. Parker and Son, 1858).

— *Novum Organon Renovatum*, 3rd edition (London: John W. Parker and Son, 1858).

— *On the Philosophy of Discovery* (London: John W. Parker and Son, 1860).

— *The Philosophy of the Inductive Sciences*, 2 vols (London: John W. Parker and Son, 1847).

Wilson, Edward O., *Consilience: The Unity of Knowledge* (New York: Random House, 1998).

— *Sociobiology*, Abridged edition (Cambridge MA: Harvard University Press, 1980).

Winegar, Reed, 'Kant's Criticisms of Hume's Dialogues Concerning Natural Religion', *British Journal for the History of Philosophy*, 23/5 (2015), 888–910.

Winkler, Kenneth, 'The New Hume', in Rupert Read and Kenneth Richman (eds), *The New Hume Debate*, Revised edition (London: Routledge, 2007), pp. 52–87.

Wood, Allen, Paul Guyer and Henry E. Allison, 'Debating Allison on Transcendental Idealism', *Kantian Review*, 12/2 (2007), 1–39.

Wood, Allen W., 'Kant's Compatibilism', in Allen W. Wood (eds), *Self and Nature in Kant's Philosophy* (London: Cornell University Press, 1984), pp. 73–101.

Yeo, Richard, 'William Whewell, Natural Theology and the Philosophy of Science in Mid Nineteenth Century Britain', *Annals of Science*, 36/5 (1979), 493–516.

Zammito, John, 'Teleology Then and Now: The Question of Kant's Relevance for Contemporary Controversies over Function in Biology', *Studies in History and Philosophy of Science Part C: Studies in History and Philosophy of Biological and Biomedical Sciences*, 37/4 (2006), 748–70.

— 'The Lenoir Thesis Revisited: Blumenbach and Kant', *Studies in History and Philosophy of Science Part C: Studies in History and Philosophy of Biological and Biomedical Sciences*, 43/1 (2012), 120–32.

Zuckert, Rachel, 'Empirical Scientific Investigation and the Ideas of Reason', in Angela Breitenbach and Michela Massimi (eds), *Kant and the Laws of Nature* (Cambridge: Cambridge University Press, 2017), pp. 89–107.

Zumbach, Clark, *The Transcendent Science: Kant's Conception of Biological Methodology*, Nijhoff International Philosophy Series (The Hague: Martinus Nijhoff Publishers, 1984).

Index